Techniques of Integration

Calculus Practice Workbook with Full Solutions

$$\int_{x=\frac{\pi}{6}}^{\frac{\pi}{3}} \sec^3 x \, dx$$

Chris McMullen, Ph.D.

Techniques of Integration
Calculus Practice Workbook
with Full Solutions
Chris McMullen, Ph.D.

Copyright © 2025 Chris McMullen, Ph.D.

www.improveyourmathfluency.com
www.monkeyphysicsblog.wordpress.com
www.chrismcmullen.com

All rights are reserved. However, educators or parents who purchase one copy of this workbook (or who borrow one physical copy from a library) may make and distribute photocopies of selected pages for instructional (non-commercial) purposes for their own students or immediate family members only.

Zishka Publishing
ISBN: 978-1-941691-53-3

Mathematics > Calculus

Contents

Introduction	iv
1 Basic Antiderivatives	5
2 Algebraic Substitutions	18
3 Trigonometric Substitutions	33
4 Logarithmic and Power Substitutions	53
5 Completing the Square	62
6 Partial Fractions	76
7 Integration by Parts	92
8 Odd/even Functions over Symmetric Limits	111
9 Improper Integrals	118
10 Double and Triple Integrals	129
Answers with Full Solutions	144

Introduction

This book covers techniques of integration that are particularly common and helpful in applications of calculus.

Prerequisite: The student should have already completed one semester of calculus and should therefore already know how to find derivatives of a variety of functions, including polynomials, trig functions, logarithms, and exponentials. (Students should already know precalculus, yet Chapter 3 will remind you of a variety of trig identities and Chapter 4 will remind you about logarithms and hyperbolic functions. Although you should have learned the derivatives of inverse trig functions in first-semester calculus, you can find a list of these derivatives in Chapter 7.) One of the main reasons that integration is difficult for many students is that it really helps to be fluent with algebra, trigonometry, and logarithms. If you aren't fluent with algebra, trigonometry, and logarithms, brushing up on the rules of algebra and the identities of precalculus may prove to be quite helpful towards mastering the art of integration.

Topics include:
- common antiderivatives that every student should know
- indefinite vs. definite integrals
- algebraic and trigonometric substitutions
- tips for making effective substitutions
- completing the square
- partial fractions
- integration by parts
- tips for choosing $u(x)$ and $v(x)$
- odd and even functions over symmetric limits
- logarithms and exponentials
- hyperbolic functions
- improper integrals
- double and triple integrals

1 Basic Antiderivatives

An **antiderivative** is the inverse of a derivative; it's basically the same operation done in reverse. This is easy to see with an example. For example, if you take a derivative of x^4 with respect to x, you get $4x^3$; the antiderivative of $4x^3$ is therefore $x^4 + c$, where c is a constant. If you begin with $4x^3$ and want to find its antiderivative, you are basically asking, "Which function can you take a derivative of and obtain $4x^3$ as a result?" The answer is $x^4 + c$, since $\frac{d}{dx}(x^4 + c) = \frac{d}{dx}x^4 + \frac{d}{dx}c = 4x^3 + 0 = 4x^3$.

Recall that the derivative of ax^b with respect to x is equal to abx^{b-1}. For example, consider $7x^4$, where $a = 7$ and $b = 4$. The derivative with respect to x is $\frac{d}{dx}7x^4 = 7(4)x^{4-1} = 28x^3$. If instead you want the antiderivative of ax^b, it is equal to $\frac{ax^{b+1}}{b+1} + c$ (if $b \neq -1$). For example, the antiderivative of $7x^4$ is $\frac{7x^{4+1}}{4+1} = \frac{7x^5}{5} + c$. You can check this by taking a derivative with respect to x, which gives $\frac{d}{dx}\left(\frac{7x^5}{5} + c\right) = \frac{7(5)x^{5-1}}{5} = 7x^4$.

There is a **valuable tip** to be found here. After you find an antiderivative, it's easy to check the answer. Just take a derivative, like we did in the previous example. As we will see in this book, it isn't always easy to find an antiderivative or to perform an integral, but it's often **easy to check the answer**; simply take a derivative.

An **indefinite integral** doesn't have any limits of integration, like the example below. For an **indefinite integral**, find the antiderivative (remember to include the constant of integration). The example below is an indefinite integral of the function $12x^5$ over the variable x; the function being integrated ($12x^5$ in this case) is called the **integrand**. The answer to the integral is the antiderivative of $12x^5$. Compare $12x^5$ with ax^b to see that $a = 12$ and $b = 5$. Since $b \neq -1$, we may use the formula $\frac{ax^{b+1}}{b+1} + c$, where c is referred to as the constant of integration.

$$\int 12x^5 \, dx = \frac{12x^{5+1}}{5+1} + c = \frac{12x^6}{6} + c = 2x^6 + c$$

Check the answer by taking a derivative: $\frac{d}{dx}(2x^6 + c) = 6(2)x^{6-1} = 12x^5$.

Chapter 1 – Basic Antiderivatives

A **definite integral** includes limits of integration. In the example below, the lower limit is 1 and the upper limit is 2. To perform a definite integral, first find the antiderivative. Then evaluate the antiderivative at the upper and lower limits and subtract. **Tip**: You can ignore the constant of integration when performing a definite integral because it will cancel out in the subtraction. To find the antiderivative in the example below, compare $6x^2$ with ax^b to see that $a = 6$ and $b = 2$. Since $b \neq -1$, use the formula $\frac{ax^{b+1}}{b+1}$.

$$\int_{x=1}^{2} 6x^2 \, dx = \left[\frac{6x^{2+1}}{2+1}\right]_{x=1}^{2} = \left[\frac{6x^3}{3}\right]_{x=1}^{2} = [2x^3]_{x=1}^{2}$$

$$= 2(2)^3 - 2(1)^3 = 2(8) - 2(1) = 16 - 2 = 14$$

In the example above, the notation $[2x^3]_{x=1}^{2}$ means to evaluate $2x^3$ at $x = 2$, also evaluate $2x^3$ at $x = 1$, and subtract. The final answer is just a number: in this case, 14.

Recall that a derivative of a function visually represents the **slope** of the tangent line. A definite integral can also be interpreted visually; it represents the **area** between the curve and the horizontal axis.

In this chapter, we will focus on the simplest antiderivatives, for which the answer can be determined readily just by recalling the formulas for common derivatives. In the remaining chapters, we will explore a variety of methods for performing integrals when the antiderivative isn't so simple.

Following are some simple formulas for antiderivatives. For example, the formulas $\int \cos x \, dx = \sin x + c$ and $\int \sin x \, dx = -\cos x + c$ follow from $\frac{d}{dx} \sin x = \cos x$ and $\frac{d}{dx} \cos x = -\sin x$. In contrast, it's not as easy to figure out $\int \sec x \, dx$ because most students don't know (without first seeing it and then memorizing it) which function you can take a derivative of and obtain secant as a result (we'll find out in Chapter 3).

$$\int ax^b \, dx = \frac{ax^{b+1}}{b+1} + c \quad (\text{if } b \neq -1) \quad , \quad \int \cos x \, dx = \sin x + c$$

$$\int ax^{-1} \, dx = \int \frac{a \, dx}{x} = a \ln|x| + c \quad (\text{if } x \neq 0) \quad , \quad \int \sin x \, dx = -\cos x + c$$

$$\int a \, dx = ax + c \quad , \quad \int e^x \, dx = e^x + c \quad , \quad \int e^{-x} \, dx = -e^{-x} + c$$

Recall the definition of the **hyperbolic** cosine and sine functions (not to be confused with the ordinary trig functions). The equations for these functions are given below.

$$\cosh x = \frac{e^x + e^{-x}}{2} \quad , \quad \sinh x = \frac{e^x - e^{-x}}{2}$$

Since $\frac{d}{dx}\cosh x = \sinh x$ and $\frac{d}{dx}\sinh x = \cosh x$, it follows that

$$\int \cosh x\, dx = \sinh x + c \quad , \quad \int \sinh x\, dx = \cosh x + c$$

Compare the antiderivatives for the hyperbolic cosine and sine functions to the antiderivatives for ordinary cosine and sine; the difference is that the antiderivative of $\sin x$ is $-\cos x + c$, whereas the antiderivative of $\sinh x$ is $\cosh x + c$. It pays to look closely because the presence of the 'h' in $\sinh x$ makes a big difference.

Occasionally, an integral is easy if you happen to recognize that the integrand is the derivative of a common function. For example, $\int \sec^2 x\, dx$ is pretty easy (whereas $\int \sec x\, dx$ is not at all straightforward; see Chapter 3). Since $\frac{d}{dx}\tan x = \sec^2 x$, it follows that $\int \sec^2 x\, dx = \tan x + c$.

The following rules of integration often come in handy.

$$\int af(x)\, dx = a\int f(x)\, dx \quad , \quad \int [f(x) + g(x)]\, dx = \int f(x)\, dx + \int g(x)\, dx$$

The first rule tells us that a constant may be factored out of an integral. For example, $\int 5\cos x\, dx$ is 5 times $\int \cos x\, dx$. If an integrand contains multiple terms (separated by plus or minus signs), it may be separated into multiple integrals according to the second rule. For example, $\int (x^2 - 3x + 4)\, dx = \int x^2\, dx - \int 3x\, dx + \int 4\, dx$.

Chapter 1 – Basic Antiderivatives

Example 1. Find the solution to the indefinite integral below.

$$\int (6x^2 - 5)\, dx$$

Solution: First, separate this into two separate integrals. We are applying the rule $\int [f(x) + g(x)]\, dx = \int f(x)\, dx + \int g(x)\, dx$ where $f(x) = 6x^2$ and $g(x) = -5$.

$$\int (6x^2 - 5)\, dx = \int 6x^2\, dx - \int 5\, dx$$

For the first integral, compare $6x^2$ with ax^b to see that $a = 6$ and $b = 2$. Since $b \neq -1$, use the formula $\frac{ax^{b+1}}{b+1}$ + a constant. For the second integral, the integrand is a constant, for which the formula is $-\int 5\, dx = -5x$ + a constant. Each indefinite integral has a constant of integration; we will combine these two constants of integration into a single constant of integration. (If you call the two constants of integration c_1 and c_2, we're basically saying that we can define a new constant c such that $c = c_1 + c_2$. It's simpler to work with c than to work with $c_1 + c_2$. It also doesn't matter whether you call the constant c, $-k$, $c_1 + c_2$, $k_1 - k_2$, etc. These are all just constants, and they are related by arithmetic. For example, $c = -k$. The convention is to collect all the constants of integration together and call the collective constants $+\,c$.)

$$\int 6x^2\, dx - \int 5\, dx = \frac{6x^{2+1}}{2+1} - 5x + c = \frac{6x^3}{3} - 5x + c = 2x^3 - 5x + c$$

Check the answer: Take a derivative of the answer with respect to x and verify that this matches the original integrand.

$$\frac{d}{dx}(2x^3 - 5x + c) = \frac{d}{dx}2x^3 - \frac{d}{dx}5x + \frac{d}{dx}c = 6x^2 - 5 + 0 = 6x^2 - 5$$

Example 2. Perform the definite integral below.

$$\int_{t=1}^{9} \sqrt{t}\, dt$$

Solution: First find the antiderivative of $\sqrt{t}$. The trick to dealing with the square root is to recall the rule $\sqrt{t} = t^{1/2}$ from algebra. (A great variety of identities from algebra and trigonometry are particularly helpful with integration. If you repeatedly find yourself stumped and later realize that a forgotten identity had been holding you back, it may be worthwhile to invest some time reviewing and practicing identities from algebra, trigonometry, exponents, and logarithms.) Once you realize that $\sqrt{t} = t^{1/2}$, compare $t^{1/2}$ with at^b to see that $a = 1$ and $b = \frac{1}{2}$. Since $b \neq -1$, use the formula $\frac{at^{b+1}}{b+1}$ + a constant. (Observe that in this example the variable is t rather than x.)

$$\int_{t=1}^{9} \sqrt{t}\, dt = \int_{t=1}^{9} t^{1/2}\, dt = \left[\frac{t^{\frac{1}{2}+1}}{\frac{1}{2}+1}\right]_{t=1}^{9} = \left[\frac{t^{3/2}}{3/2}\right]_{t=1}^{9} = \left[\frac{2}{3}t^{3/2}\right]_{t=1}^{9}$$

The brackets [ ] indicate that $\frac{2}{3}t^{3/2}$ is just the antiderivative of $\sqrt{t}$ and that we still need to evaluate the antiderivative over the limits (from $t = 1$ to 9) since this is a definite integral rather than an indefinite integral. Note that $\frac{1}{2} + 1 = \frac{1}{2} + \frac{2}{2} = \frac{3}{2}$ and that $\frac{1}{3/2} = \frac{2}{3}$. (If you're rusty with how to find a common denominator, add fractions, divide fractions, work with reciprocals, etc., it may be worthwhile to practice these essential fraction skills because they will come in handy. Many students who struggle with calculus do so because they aren't fluent in fractions, algebra identities, and trig identities. Calculus is hard enough when you're fluent with the prerequisites; if you're not fluent with these skills, you're making calculus twice as hard as it should be. Review and practice is the remedy.) In the last expression above, we will first replace t with the upper limit, which is 9, and then the lower limit, which is 1, and we will subtract these two terms.

$$\left[\frac{2}{3}t^{3/2}\right]_{t=1}^{9} = \frac{2}{3}(9)^{3/2} - \frac{2}{3}(1)^{3/2}$$

You can do this with a calculator or by hand. For example, a calculator will tell you that $9^{3/2} = 27$. Without a calculator, you can use the rule $x^m x^n = x^{m+n}$ to see that $9^{3/2} = 9^1 9^{1/2} = 9\sqrt{9} = 9(3) = 27$ (since $1 + \frac{1}{2} = \frac{3}{2}$).

$$\frac{2}{3}(9)^{3/2} - \frac{2}{3}(1)^{3/2} = \frac{2}{3}(27) - \frac{2}{3}(1) = \frac{54 - 2}{3} = \frac{52}{3}$$

The answer to a definite integral is a single numerical value, whereas the answer to an indefinite integral is an algebraic expression. Compare the answers to Examples 1-2. Check the antiderivative: Take a derivative of $\frac{2}{3}t^{3/2}$ with respect to t.

$$\frac{d}{dt}\frac{2}{3}t^{3/2} = \frac{2}{3}\left(\frac{3}{2}\right)t^{1/2} = t^{1/2} = \sqrt{t}$$

Chapter 1 – Basic Antiderivatives

Example 3. Find the solution to the indefinite integral below, where $u \neq 0$.

$$\int \frac{3}{u} du$$

Solution: It's worth memorizing that $\int au^{-1} du = \int \frac{a\, du}{u} = a \ln|u| + c$ because this integral arises in many common applications of calculus. In this example, $a = 3$. One way to remember the absolute values is to recall from precalculus that a logarithm is only defined and real if the argument is positive.

$$\int \frac{3}{u} du = 3 \ln|u| + c$$

Alternatively, the answer may be expressed as

$$\int \frac{3}{u} du = 3 \ln|ku|$$

using k to represent the constant of integration. We'll discuss this alternative answer after checking the answer.

Check the answer: If u is positive, we get

$$\frac{d}{du}(3 \ln u + c) = \frac{d}{du} 3 \ln u + \frac{d}{du} c = \frac{3}{u} + 0 = \frac{3}{u}$$

On the other hand, if u is negative, it's a little trickier. First, note that $\frac{d}{du} \ln(ku) = \frac{1}{u}$ if k is a constant. Read this carefully to note how k doesn't appear on the right-hand side. If you've forgotten this subtle yet important point, there are two different ways to convince yourself that it's correct. It's unlike other functions. For example, compare $\frac{d}{du} \sin(ku) = k \cos(ku)$ with the above rule; with sine, the k does appear on the right-hand side. (If you've forgotten this point, you should take some time to review your derivatives. Performing integrals is really challenging if you struggle with the rules of differentiation.) If you recall the rule $\ln(yu) = \ln y + \ln u$, it's easy to understand the above rule. Letting $y = k$, we get $\frac{d}{du} \ln(ku) = \frac{d}{du} (\ln k + \ln u) = 0 + \frac{d}{du} \ln u = \frac{1}{u}$. Here, we see that the k doesn't show up on the right-hand side because the derivative of a constant is zero (and $\ln k$ is a constant because k is a constant). The other way to see the above rule is to apply the chain rule, which states that $\frac{df}{du} = \frac{df}{dg}\frac{dg}{du}$. Let $f = \ln g$ and $g = ku$ to get $\frac{d}{du} \ln(ku) = \frac{df}{du} = \frac{df}{dg}\frac{dg}{du} = \frac{d}{dg} \ln g \frac{d}{du} ku = \frac{1}{g} k = \frac{1}{ku} k = \frac{1}{u}$. Here, the k isn't on the right-hand side because it cancels out. Either way leads to $\frac{d}{du}\ln(ku) = \frac{1}{u}$. Now let $k = -1$ to see that $\frac{d}{du}\ln(-u) = \frac{1}{u}$. First, note that the right-hand side is positive, not negative, for the same reason that k doesn't show up on the right-hand side. This is very important. Secondly, $\ln(-u)$ is only real if u is negative (so that $-u$ is positive). Now we're prepared to work out the case where u is negative. When u is negative, the quantity $-u$ is positive. In this case, $\ln|u| = \ln(-u)$. In this case, we get the same as before. As explained in the previous paragraph, the minus sign doesn't show up on the right-hand side. It's a common mistake for students to think that $\frac{d}{du}\ln(-u)$ should include a minus sign before the $\frac{1}{u}$, but it's positive. If necessary, work through the previous paragraph again; the right-hand side is positive. The absolute values in the answer to this example, $3\ln|u| + c$, are important, and allow for both the case where u is positive and where u is negative, and in each case the derivative of $3\ln|u| + c$ with respect to u is a positive $\frac{3}{u}$. It's important to remember these absolute values and to also understand them. Many students who pay attention remember the absolute values, but not as many understand them, and it's tricky to remember that in the case where u is negative, checking the answer with a derivative is still positive. But if you understand the explanation in the previous paragraph, then you're equipped to do it.

$$\frac{d}{du}[3\ln(-u) + c] = \frac{d}{du}3\ln(-u) + \frac{d}{du}c = \frac{3}{u} + 0 = \frac{3}{u}$$

About the alternative answer: If you understood the previous paragraphs, you should understand the alternative answer, as they're related, but we'll examine it a different way here. The answers $3\ln|u| + c$ and $3\ln|ku|$ are equivalent. One way to see this is to apply the identity $\ln(yu) = \ln y + \ln u$ with $y = k$. This gives

$$3\ln(ku) = 3\ln k + 3\ln u$$

Observe that $3\ln k$ is a constant since 3 and k are constants. If you let $c = 3\ln k$, then $3\ln|ku|$ will be identical to $3\ln|u| + c$.

Chapter 1 – Basic Antiderivatives

Example 4. Perform the definite integral below, where the angles are in radians.

$$\int_{\theta=0}^{\pi/6} 4\cos\theta\, d\theta$$

Solution: The antiderivative of cosine is sine (since a derivative of sine equals cosine).

$$\int_{\theta=0}^{\pi/6} 4\cos\theta\, d\theta = 4\int_{\theta=0}^{\pi/6} \cos\theta\, d\theta = 4[\sin\theta]_{\theta=0}^{\pi/6}$$

For a definite integral, evaluate the antiderivative at the limits and subtract.

$$4[\sin\theta]_{\theta=0}^{\pi/6} = 4\sin\frac{\pi}{6} - 4\sin 0 = 4\left(\frac{1}{2}\right) - 4(0) = 2 - 0 = 2$$

Notes: Recall that $\frac{\pi}{6}$ radians equates to 30°. (The conversion factor is π rad = 180°.)
The sine of $\frac{\pi}{6}$ radians (or 30°) equals one-half.

Check the antiderivative: Take a derivative of $4\sin\theta$ with respect to θ and compare with the original integrand.

$$\frac{d}{d\theta}4\sin\theta = 4\cos\theta$$

Example 5. Find the solution to the indefinite integral below.

$$\int \cosh x\, dx$$

Solution: The antiderivative of hyperbolic cosine is hyperbolic sine. (Note that $\cosh x$ is different from the ordinary trig function: $\cosh x = \frac{e^x + e^{-x}}{2}$.)

$$\int \cosh x\, dx = \sinh x + c$$

Check the answer: The derivative of hyperbolic sine is hyperbolic cosine. (Recall that $\sinh x = \frac{e^x - e^{-x}}{2}$.)

$$\frac{d}{dx}(\sinh x + c) = \frac{d}{dx}\sinh x + \frac{d}{dx}c = \cosh x + 0 = \cosh x$$

Techniques of Integration Calculus Practice Workbook

Example 6. Perform the definite integral below.
$$\int_{t=0}^{1} e^{-t}\, dt$$

Solution: The antiderivative of e^{-t} equals $-e^{-t}$. Be careful with the minus signs.

$$\int_{t=0}^{1} e^{-t}\, dt = [-e^{-t}]_{t=0}^{1} = -e^{-1} - (-e^{-0}) = -\frac{1}{e} + 1 = \frac{-1+e}{e} = \frac{e-1}{e} \approx 0.63212$$

Sign check: Recall that a definite integral represents the area between the curve and the t-axis. If you graph e^{-t}, the area is positive because e^{-t} lies above the t-axis. Check the antiderivative: Take a derivative of $-e^{-t}$ with respect to t and compare with the original integrand.

$$\frac{d}{dt}(-e^{-t}) = -(-e^{-t}) = e^{-t}$$

Example 7. Find the solution to the indefinite integral below.

$$\int \sec\theta \tan\theta\, d\theta$$

Solution: This should be easy if you're fluent with your trig derivatives. You should recall from first-semester calculus that $\frac{d}{d\theta}\sec\theta = \sec\theta\tan\theta$.

$$\int \sec\theta \tan\theta\, d\theta = \sec\theta + c$$

Check the answer: Take a derivative of $\sec\theta$ with respect to θ.

$$\frac{d}{d\theta}(\sec\theta + c) = \frac{d}{d\theta}\sec\theta + \frac{d}{d\theta}c = \sec\theta\tan\theta + 0 = \sec\theta\tan\theta$$

Another way to see this is to recall that $\sec\theta = \frac{1}{\cos\theta}$ and $\tan\theta = \frac{\sin\theta}{\cos\theta}$.

$$\frac{d}{d\theta}\sec\theta = \frac{d}{d\theta}\frac{1}{\cos\theta}$$

Apply the chain rule with $f = \frac{1}{u}$ and $u = \cos\theta$. The chain rule says that $\frac{df}{d\theta} = \frac{df}{du}\frac{du}{d\theta}$. (If you forgot the chain rule from first-semester calculus, it is worth reviewing.)

$$\frac{d}{d\theta}\frac{1}{\cos\theta} = \frac{df}{d\theta} = \frac{df}{du}\frac{du}{d\theta} = \frac{d}{du}\frac{1}{u}\frac{d}{d\theta}\cos\theta = -\frac{1}{u^2}(-\sin\theta)$$

$$= \frac{\sin\theta}{\cos^2\theta} = \frac{1}{\cos\theta}\frac{\sin\theta}{\cos\theta} = \sec\theta\tan\theta$$

Chapter 1 Problems

Directions: Perform each integral.

❶ $\displaystyle\int 56x^7\, dx$

❷ $\displaystyle\int (t^3 - t)\, dt$

❸ $\displaystyle\int_{x=2}^{3} 12x^2\, dx$

❹ $\displaystyle\int_{u=3}^{6} (u^2 - 4u + 2)\, du$

5 $\displaystyle\int x^{2/3}\,dx$

6 $\displaystyle\int \frac{dy}{\sqrt{y}}$

7 $\displaystyle\int_{t=1}^{16} t^{3/4}\,dt$

8 $\displaystyle\int_{z=3}^{24} \frac{dz}{z}$

Chapter 1 – Basic Antiderivatives

9 $\int (\cos\theta - \sin\theta)\, d\theta$

10 $\int (\cosh x - \sinh x)\, dx$

11 $\displaystyle\int_{\theta=\pi/6}^{\pi/4} \sin\theta\, d\theta$

12 $\displaystyle\int_{x=0}^{1} \frac{e^x - 1}{e^x}\, dx$

⓭ $\displaystyle\int \sec^2\theta\, d\theta$

⓮ $\displaystyle\int \frac{\cos x}{\sin^2 x}\, dx$

⓯ $\displaystyle\int \frac{d\varphi}{\sin^2\varphi}$

2 Algebraic Substitutions

Sometimes an integral can be made much simpler just by changing variables. If the given integral is a function of x, it may be possible to rewrite it as a function of a new variable u, where the new integral is simpler. But there's a catch. To do this correctly, you need to make **two** substitutions: first rewrite x in terms of u, and also rewrite the differential element dx in terms of du. This is easiest to understand with an example.

Consider the integral $\int x^2 \sqrt{x^3 + 4}\, dx$. The first substitution $u = x^3 + 4$ makes the radical simpler. That is, $\sqrt{x^3 + 4}$ becomes $\sqrt{u}$. But that's only one of two substitutions that we must make. To figure out the second substitution, take a **derivative** of $u = x^3 + 4$ with respect to x to get $\frac{du}{dx} = 3x^2$, which we can write as $du = 3x^2 dx$. Examine the given integral. We'll use $u = x^3 + 4$ to rewrite the radical $\sqrt{x^3 + 4}$ as $\sqrt{u}$, and we'll replace $x^2 dx$ with $\frac{du}{3}$ (which follows from $du = 3x^2 dx$). After both substitutions, the integral becomes $\frac{1}{3}\int \sqrt{u}\, du$. We started with an integral involving x and finished with an integral involving u, which is simpler than the original integral. See Example 1.

Let's see if you understand the basic idea. Which of these three integrals is easy to do using the substitution method: $\int \sin(x^2)\, dx$, $\int x \sin(x^2)\, dx$, or $\int x^2 \sin(x^2)\, dx$? If you get the main idea, the answer is the middle integral. Why? Because the derivative of x^2 equals $2x$, which includes x to the first power, like the middle integral. To see this, let $u = x^2$. The derivative is $\frac{du}{dx} = 2x$, for which $du = 2xdx$, which we can rewrite as $\frac{du}{2} = xdx$. When we replace x^2 with u and also replace xdx with $\frac{du}{2}$, the middle integral becomes $\frac{1}{2}\int \sin u\, du$. See Example 2.

Side note: This technique is called the **substitution rule**. It basically applies the **chain rule** (of derivatives) to integrals. (You may recall that we applied the chain rule in Examples 3 and 7 in Chapter 1. If you've forgotten the chain rule from first-semester calculus, it would be worth reviewing it.)

Tips for making effective substitutions:

- **Look for a derivative**. It's a common mistake to only think about how to rewrite x in terms of u, ignoring how to rewrite dx in terms of du. Students who excel at the substitution method think about the derivative $\frac{du}{dx}$ and look for it in the given integral. For example, consider $\int x^2(x^3 - 1)^8 \, dx$. The factor of x^2 makes this easy. With $u = x^3 - 1$, the derivative is $\frac{du}{dx} = 3x^2$. See Example 3.
- If the substitution is **linear**, like $u = 2x - 5$, then the derivative is a **constant**. Here, $\frac{du}{dx} = 2$, such that $du = 2dx$ or $\frac{du}{2} = dx$. This derivative doesn't have a variable; it just has a constant. You don't have to 'look' for this derivative; it comes naturally. For example, $\int \frac{dx}{2x-5}$ turns into $\frac{1}{2} \int \frac{du}{u}$. See Example 4.
- What will **make the integrand appear simpler**? For example, $\int \frac{dx}{\sqrt{3x+2}}$ would be simpler if $3x + 2$ were a single variable. Let $u = 3x + 2$, such that $du = 3dx$ or $\frac{du}{3} = dx$. This gives $\frac{1}{3} \int \frac{du}{\sqrt{u}}$. See Example 5. (You might wonder why we don't let u equal $\sqrt{3x + 2}$. The reason is that $\frac{du}{dx}$ is way more complicated in that case.)
- A few of the trickier integrals that can be done by substitution involve some algebra (or trig) skills, such as **factoring**. For example, $\int \frac{x}{x^4+2x^2+1} \, dx$ is simpler if you rewrite the denominator as $\int \frac{x}{(x^2+1)^2} \, dx$, and $\int x^3\sqrt{x^2 + 4} \, dx$ is simpler if you rewrite it as $\int x^2 x \sqrt{x^2 + 4} \, dx$. See Examples 6-7.
- If an integral involves $x^2 + c$, $x^2 - c$, or $c - x^2$, where c is a constant, if you don't see an x in the integrand that could help form du (recall the first bullet point above), use a **trig substitution** (Chapter 3). For example, $\int \frac{dx}{x^2+9}$ can be done using a trig substitution. In contrast, $\int \frac{x}{x^2+9} \, dx$ is easy if you let $u = x^2 + 9$ because $\frac{du}{dx} = 2x$ includes the x in the integrand. See Example 8.
- Don't be afraid to use "trial and error." It's a common experience for the first substitution that comes to mind to not work out. If a substitution doesn't work, if you can figure out why it didn't work, it may help to find a better substitution. Or try a different approach. Students who give up never solve the problem.

Chapter 2 – Algebraic Substitutions

For a **definite integral**, use the equation for the substitution to obtain the new limits of integration. For example, suppose that the original integral varies from $x = 1$ to $x = 4$ and you have made the substitution $u = x^2 - 1$, for which $du = 2xdx$. Plug the given lower and upper limits into $u = x^2 - 1$ to find that the new limits of integration are $u = 1^2 - 1 = 1 - 1 = 0$ and $u = 4^2 - 1 = 16 - 1 = 15$. Examples 2, 4, 6, and 8 will illustrate this further.

Example 1. Find the solution to the indefinite integral below.
$$\int x^2 \sqrt{x^3 + 4}\, dx$$
Solution: Since the derivative of $x^3 + 4$ with respect to x equals $3x^2$ and there happens to be an x^2 in the integrand, this suggests the substitution $u = x^3 + 4$. A derivative of u with respect to x equals $\frac{du}{dx} = \frac{d}{dx}(x^3 + 4) = 3x^2$, which we may rewrite as $du = 3x^2 dx$, which equates to $\frac{du}{3} = x^2 dx$. Replace $x^3 + 4$ with u and replace $x^2 dx$ with $\frac{du}{3}$.
$$\int x^2 \sqrt{x^3 + 4}\, dx = \int \frac{\sqrt{u}}{3}\, du = \frac{1}{3}\int \sqrt{u}\, du$$
This integral is like Example 2 from Chapter 1. Use the identity $\sqrt{u} = u^{1/2}$. Apply the formula $\frac{au^{b+1}}{b+1} + c$ with $a = \frac{1}{3}$ and $b = \frac{1}{2}$. Note that $\frac{1}{3/2} = \frac{2}{3}$.
$$\frac{1}{3}\int \sqrt{u}\, du = \frac{1}{3}\int u^{1/2}\, du = \frac{1}{3}\left(\frac{u^{\frac{1}{2}+1}}{\frac{1}{2}+1}\right) + c = \frac{1}{3}\frac{u^{3/2}}{3/2} + c = \frac{1}{3}\frac{2u^{3/2}}{3} + c = \frac{2}{9}u^{3/2} + c$$
We're not finished yet because our current answer involves u instead of x. Use the equation $u = x^3 + 4$ to express the final answer in terms of x.
$$\int x^2 \sqrt{x^3 + 4}\, dx = \frac{2(x^3 + 4)^{3/2}}{9} + c$$
Check the answer: Take a derivative of the answer with respect to x and verify that this matches the original integrand. Apply the chain rule (Chapter 1, Examples 3 and 7) with $f = \frac{2u^{3/2}}{9} + c = \frac{2(x^3+4)^{3/2}}{9} + c$ and $u = x^3 + 4$.
$$\frac{df}{dx} = \frac{df}{du}\frac{du}{dx} = \frac{d}{du}\left(\frac{2u^{3/2}}{9} + c\right)\frac{d}{dx}(x^3 + 4) = \frac{3}{2}\frac{2}{9}(x^3 + 4)^{1/2}(3x^2) = x^2\sqrt{x^3 + 4}$$

Techniques of Integration Calculus Practice Workbook

Example 2. Perform the definite integral below. Note: $\sin(\theta^2)$ means to square θ first and then take the sine, whereas $\sin^2 \theta$ means to first evaluate the sine at θ and then square that value.

$$\int_{\theta=0}^{\sqrt{\pi}} \theta \sin(\theta^2) \, d\theta$$

Solution: Since the derivative of θ^2 with respect to θ equals 2θ and there happens to be a θ in the integrand, this suggests the substitution $u = \theta^2$. A derivative of u with respect to θ equals $\frac{du}{d\theta} = \frac{d}{d\theta}\theta^2 = 2\theta$, which we may rewrite as $du = 2\theta d\theta$, which equates to $\frac{du}{2} = \theta d\theta$. Replace θ^2 with u and replace $\theta d\theta$ with $\frac{du}{2}$. Plug the given limits (0 and $\sqrt{\pi}$) into the equation $u = \theta^2$ to find the new limits of integration: $0^2 = 0$ and $\left(\sqrt{\pi}\right)^2 = \pi$. **Remember to change the limits of integration in the new integral.**

$$\int_{\theta=0}^{\sqrt{\pi}} \theta \sin(\theta^2) \, d\theta = \int_{u=0}^{\pi} \frac{\sin u}{2} du = \frac{1}{2}\int_{u=0}^{\pi} \sin u \, du = \frac{[-\cos u]_{u=0}^{\pi}}{2}$$

$$= -\frac{\cos \pi}{2} - \left(-\frac{\cos 0}{2}\right) = -\frac{\cos \pi}{2} + \frac{\cos 0}{2} = -\left(-\frac{1}{2}\right) + \frac{1}{2} = \frac{1}{2} + \frac{1}{2} = 1$$

Notes: The variable u is in radians,[1] where π radians equate to 180°. A definite integral equals the area between the curve and the horizontal axis. Knowing this helps check the signs. Since the sine function is positive in Quadrants I and II, the area under the curve is positive from 0 to π, so the integral must be positive. (If you make a mistake with the signs and get a negative answer or zero, that obviously can't be right.)

Check the antiderivative: Take a derivative of $-\frac{\cos u}{2} = -\frac{\cos(\theta^2)}{2}$ with respect to θ.

Apply the chain rule (Example 1) with $f = -\frac{\cos u}{2} = -\frac{\cos(\theta^2)}{2}$ and $u = \theta^2$.

$$\frac{df}{d\theta} = \frac{df}{du}\frac{du}{d\theta} = \frac{d}{du}\left(-\frac{\cos u}{2}\right)\frac{d}{d\theta}\theta^2 = \left(\frac{\sin u}{2}\right)(2\theta) = \theta \sin(\theta^2)$$

---

[1] The original argument of sine has θ^2. Since you take the sine of an angle, in this example θ^2 is an angle, whereas θ is the square root of an angle. In this example, θ^2 has units of radians (but θ doesn't).

Chapter 2 – Algebraic Substitutions

Example 3. Find the solution to the indefinite integral below.
$$\int t^2(t^3-1)^8\,dt$$
Solution: Since the derivative of t^3-1 with respect to t equals $3t^2$ and there happens to be a t^2 in the integrand, this suggests the substitution $u = t^3 - 1$. A derivative of u with respect to t equals $\frac{du}{dt} = \frac{d}{dt}(t^3-1) = 3t^2$, which we may rewrite as $du = 3t^2 dt$, which equates to $\frac{du}{3} = t^2 dt$. Replace t^3-1 with u and replace $t^2 dt$ with $\frac{du}{3}$. Then, in the answer, replace u with t^3-1.
$$\int t^2(t^3-1)^8\,dt = \int \frac{u^8}{3}du = \frac{1}{3}\int u^8\,du = \frac{1}{3}\left(\frac{u^9}{9}\right) + c = \frac{u^9}{27} + c = \frac{(t^3-1)^9}{27} + c$$
Check the answer: Take a derivative of the answer with respect to x and verify that this matches the original integrand. Apply the chain rule (Example 1) with $f = \frac{u^9}{27} + c$
$$= \frac{(t^3-1)^9}{27} + c \text{ and } u = t^3 - 1.$$
$$\frac{df}{dt} = \frac{df}{du}\frac{du}{dt} = \frac{d}{du}\left(\frac{u^9}{27} + c\right)\frac{d}{dt}(t^3-1) = \frac{9u^8}{27}(3t^2) = (t^3-1)^8 t^2 = t^2(t^3-1)^8$$

Example 4. Perform the definite integral below.
$$\int_{x=3}^{6}\frac{dx}{2x-5}$$
Solution: The only expression in this integrand is $2x - 5$. Since a derivative of $2x - 5$ with respect to x is a constant (equal to 2) instead of a variable, we don't need to 'look' for a derivative in this case. As we'll see, the substitution from dx to du will naturally take care of this constant. Let $u = 2x - 5$. A derivative of u with respect to x equals $\frac{du}{dx} = \frac{d}{dx}(2x-5) = 2$, which we may rewrite as $du = 2dx$, which equates to $\frac{du}{2} = dx$. Replace $2x - 5$ with u and replace dx with $\frac{du}{2}$. Plug the given limits (3 and 6) into the equation $u = 2x - 5$ to find the new limits of integration: $2(3) - 5 = 6 - 5 = 1$ and $2(6) - 5 = 12 - 5 = 7$. When applying the substitution method to a definite integral, **don't forget to change the limits of integration in the new integral.**

$$\int_{x=3}^{6}\frac{dx}{2x-5} = \int_{u=1}^{7}\frac{du}{2u} = \frac{1}{2}\int_{u=1}^{7}\frac{du}{u}$$

This integral is like Example 3 from Chapter 1.

$$\frac{1}{2}\int_{u=1}^{7}\frac{du}{u} = \frac{1}{2}[\ln|u|]_{u=1}^{7} = \frac{1}{2}(\ln 7 - \ln 1) = \frac{1}{2}\ln\frac{7}{1} = \frac{1}{2}\ln 7 \approx 0.973$$

Note: We used the logarithm identity $\ln p - \ln q = \ln\frac{p}{q}$.

Check the antiderivative: Take a derivative of $\frac{1}{2}\ln u = \frac{1}{2}\ln(2x-5)$ with respect to x.

Apply the chain rule (Example 1) with $f = \frac{1}{2}\ln u = \frac{1}{2}\ln(2x-5)$ and $u = 2x - 5$.

$$\frac{df}{dx} = \frac{df}{du}\frac{du}{dx} = \frac{d}{du}\left(\frac{1}{2}\ln u\right)\frac{d}{dx}(2x-5) = \left(\frac{1}{2u}\right)(2) = \frac{1}{u} = \frac{1}{2x-5}$$

Example 5. Find the solution to the indefinite integral below.

$$\int \frac{dy}{\sqrt{3y+2}}$$

Solution: Similar to Example 4, here the only expression is $3y + 2$, and it's a linear function (since the variable is raised to the first power), so the derivative will be a constant. Let $u = 3y + 2$. A derivative of u with respect to y equals $\frac{du}{dy} = \frac{d}{dy}(3y+2) = 3$, which we may rewrite as $du = 3dy$, which equates to $\frac{du}{3} = dy$. Replace $3y + 2$ with u and replace dy with $\frac{du}{3}$.

$$\int \frac{dy}{\sqrt{3y+2}} = \int \frac{du}{3\sqrt{u}} = \frac{1}{3}\int \frac{du}{\sqrt{u}}$$

This is like Problem 6 from Chapter 1. Recall from algebra that $\frac{1}{\sqrt{u}} = \frac{1}{u^{1/2}} = u^{-1/2}$.

Apply the formula $\frac{au^{b+1}}{b+1} + c$ with $a = \frac{1}{3}$ and $b = -\frac{1}{2}$. Note that $\frac{1}{1/2} = 2$.

$$\frac{1}{3}\int \frac{du}{\sqrt{u}} = \frac{1}{3}\int u^{-1/2}\,du = \frac{1}{3}\left(\frac{u^{-\frac{1}{2}+1}}{-\frac{1}{2}+1}\right) + c = \frac{1}{3}\frac{u^{1/2}}{1/2} + c = \frac{2}{3}\sqrt{u} + c = \frac{2}{3}\sqrt{3y+2} + c$$

23

Chapter 2 – Algebraic Substitutions

Check the answer: Take a derivative of the answer with respect to y and verify that this matches the original integrand. Apply the chain rule (Example 1) with $f = \frac{2}{3}\sqrt{u} + c = \frac{2}{3}\sqrt{3y+2} + c$ and $u = 3y + 2$. Note that $u^{-1/2} = \frac{1}{u^{1/2}} = \frac{1}{\sqrt{u}}$.

$$\frac{df}{dy} = \frac{df}{du}\frac{du}{dy} = \frac{d}{du}\left(\frac{2}{3}\sqrt{u} + c\right)\frac{d}{dy}(3y+2) = \frac{d}{du}\left(\frac{2}{3}u^{1/2} + c\right)(3)$$

$$= \left(\frac{2}{3}\frac{1}{2}u^{-1/2}\right)(3) = \frac{1}{u^{1/2}} = \frac{1}{\sqrt{u}} = \frac{1}{\sqrt{3y+2}}$$

Example 6. Perform the definite integral below.

$$\int_{x=0}^{2} \frac{x}{x^4 + 2x^2 + 1} dx$$

Solution: It would be **incorrect** to define u to be $x^4 + 2x^2 + 1$ because in that case $\frac{du}{dx}$ would equal $4x^3 + 4x$; that derivative is nowhere to be found. The 'trick' here is to realize that the denominator is a perfect square; it can be **factored** using algebra as follows: $x^4 + 2x^2 + 1 = (x^2 + 1)^2$. This is easy to check; just multiply $x^2 + 1$ by itself: $(x^2 + 1)(x^2 + 1) = x^4 + 2x^2 + 1$. Let $u = x^2 + 1$. A derivative of u with respect to x equals $\frac{du}{dx} = \frac{d}{dx}(x^2 + 1) = 2x$, which we may rewrite as $du = 2xdx$, which equates to $\frac{du}{2} = xdx$. Rewrite $x^4 + 2x^2 + 1$ as $(x^2 + 1)^2$, replace $x^2 + 1$ with u, and replace xdx with $\frac{du}{2}$. Plug the given limits (0 and 2) into the equation $u = x^2 + 1$ to find the new limits of integration: $0^2 + 1 = 1$ and $2^2 + 1 = 5$. **Remember to change the limits.**

$$\int_{x=0}^{2} \frac{x}{x^4 + 2x^2 + 1} dx = \int_{x=0}^{2} \frac{x}{(x^2+1)^2} dx = \int_{u=1}^{5} \frac{du}{2u^2} = \frac{1}{2}\int_{u=1}^{5} \frac{du}{u^2}$$

Recall from algebra that $\frac{1}{u^2} = u^{-2}$. Apply the formula $\frac{au^{b+1}}{b+1} + c$ with $a = \frac{1}{2}$ and $b = -2$.

$$\frac{1}{2}\int_{u=1}^{5} \frac{du}{u^2} = \frac{1}{2}\int_{u=1}^{5} u^{-2} du = \frac{1}{2}\left[\frac{u^{-2+1}}{-2+1}\right]_{u=1}^{5} = \frac{1}{2}\left[\frac{u^{-1}}{-1}\right]_{u=1}^{5} = -\frac{1}{2}\left[\frac{1}{u}\right]_{u=1}^{5}$$

$$= -\frac{1}{2}\left(\frac{1}{5} - \frac{1}{1}\right) = -\frac{1}{2}\left(\frac{1}{5} - \frac{5}{5}\right) = -\frac{1}{2}\left(\frac{1-5}{5}\right) = -\frac{1}{2}\left(-\frac{4}{5}\right) = \frac{4}{10} = \frac{2}{5} = 0.4$$

Check the antiderivative: Take a derivative of $-\frac{1}{2}u^{-1} = -\frac{1}{2}(x^2+1)^{-1}$ with respect to x. Apply the chain rule (Example 1) with $f = -\frac{1}{2}u^{-1}$ and $u = x^2 + 1$.

$$\frac{df}{dx} = \frac{df}{du}\frac{du}{dx} = \frac{d}{du}\left(-\frac{1}{2}u^{-1}\right)\frac{d}{dx}(x^2+1) = (-1)\left(-\frac{1}{2}\right)u^{-2}(2x)$$

$$= u^{-2}x = \frac{x}{u^2} = \frac{x}{(x^2+1)^2} = \frac{x}{x^4+2x^2+1}$$

Example 7. Find the solution to the indefinite integral below.

$$\int x^3\sqrt{x^2+4}\,dx$$

Solution: A derivative of $x^2 + 4$ with respect to x equals $2x$, but unfortunately there is an x^3 in the integrand, which is a different power. The 'trick' is to factor x^3 using $x^3 = x(x^2)$ as shown below.

$$\int x^3\sqrt{x^2+4}\,dx = \int x(x^2)\sqrt{x^2+4}\,dx$$

Let $u = x^2 + 4$. A derivative of u with respect to x equals $\frac{du}{dx} = \frac{d}{dx}(x^2+4) = 2x$, which we may rewrite as $du = 2x\,dx$, which equates to $\frac{du}{2} = x\,dx$. In this example, we will make three substitutions all together (whereas we only needed two substitutions in the previous examples): replace $x^2 + 4$ with u in the radical, replace $x\,dx$ with $\frac{du}{2}$, and also replace the remaining x^2 with $u - 4$ (this comes from $u = x^2 + 4$; subtract 4 from both sides to get $u - 4 = x^2$).

$$\int x(x^2)\sqrt{x^2+4}\,dx = \frac{1}{2}\int (u-4)\sqrt{u}\,du$$

Oh, no! How will we integrate this expression? Actually, it's easier than it may look. Recall from algebra that $\sqrt{u} = u^{1/2}$. All we need to do is **distribute**.

$$\frac{1}{2}\int (u-4)\sqrt{u}\,du = \frac{1}{2}\int (u-4)u^{1/2}\,du = \frac{1}{2}\int \left(u^{3/2} - 4u^{1/2}\right)du$$

$$= \frac{1}{2}\frac{u^{\frac{3}{2}+1}}{\frac{3}{2}+1} - \frac{1}{2}(4)\frac{u^{\frac{1}{2}+1}}{\frac{1}{2}+1} + c = \frac{1}{2}\frac{u^{5/2}}{5/2} - 2\frac{u^{3/2}}{3/2} + c = \frac{1}{2}\frac{2}{5}u^{5/2} - 2\frac{2}{3}u^{3/2} + c$$

$$= \frac{u^{5/2}}{5} - \frac{4u^{3/2}}{3} + c = \frac{(x^2+4)^{5/2}}{5} - \frac{4(x^2+4)^{3/2}}{3} + c$$

Chapter 2 – Algebraic Substitutions

Check the answer: Take a derivative of the answer with respect to x and verify that this matches the original integrand. Apply the chain rule, like the previous examples.

$$\frac{d}{dx}\left[\frac{(x^2+4)^{5/2}}{5} - \frac{4(x^2+4)^{3/2}}{3} + c\right] = \frac{1}{5}\frac{5}{2}(x^2+4)^{3/2}(2x) - \frac{4}{3}\frac{3}{2}(x^2+4)^{1/2}(2x)$$

$$\frac{1}{2}(x^2+4)^{3/2}(2x) - 2(x^2+4)^{1/2}(2x) = x(x^2+4)^{3/2} - 4x(x^2+4)^{1/2}$$

This doesn't look like the integrand. Does that mean that the answer is wrong? No, we just need to do some algebra to get it into the same form as the integrand. Note that $(x^2+4)^{1/2} = \sqrt{x^2+4}$ and $(x^2+4)^{3/2} = (x^2+4)^1(x^2+4)^{1/2} = (x^2+4)\sqrt{x^2+4}$.

$$x(x^2+4)^{3/2} - 4x(x^2+4)^{1/2} = (x^2+4)x\sqrt{x^2+4} - 4x\sqrt{x^2+4}$$

Now factor out the $x\sqrt{x^2+4}$.

$$= (x^2+4-4)x\sqrt{x^2+4} = (x^2)x\sqrt{x^2+4} = x^3\sqrt{x^2+4}$$

Example 8. Perform the definite integral below.

$$\int_{t=1}^{3} \frac{t}{t^2+9}\,dt$$

Solution: Often, when an expression of the form $t^2 + 9$ appears, the best strategy is to use trig substitution, as we'll learn in Chapter 3. However, since in this case there is a t with the dt in the integrand and since a derivative of t^2 with respect to t equals $2t$, we can solve this simply with the substitution $u = t^2 + 9$. A derivative of u with respect to t equals $\frac{du}{dt} = \frac{d}{dt}(t^2+9) = 2t\,dt$, which we may rewrite as $du = 2t\,dt$, which equates to $\frac{du}{2} = t\,dt$. Replace $t^2 + 9$ with u and replace $t\,dt$ with $\frac{du}{2}$. Plug the given limits (1 and 3) into the equation $u = t^2 + 9$ to find the new limits of integration: $1^2 + 9 = 1 + 9 = 10$ and $3^2 + 9 = 9 + 9 = 18$. **Don't forget to change the limits.**

$$\int_{t=1}^{3} \frac{t}{t^2+9}\,dt = \int_{u=10}^{18} \frac{du}{2u} = \frac{1}{2}\int_{u=10}^{18} \frac{du}{u}$$

This integral is like Example 3 from Chapter 1.

$$\frac{1}{2}\int_{u=10}^{18} \frac{du}{u} = \frac{1}{2}[\ln|u|]_{u=10}^{18} = \frac{1}{2}(\ln 18 - \ln 10) = \frac{1}{2}\ln\left(\frac{18}{10}\right) = \frac{1}{2}\ln\left(\frac{9}{5}\right) = \frac{1}{2}\ln(1.8) \approx 0.294$$

Note: We used the logarithm identity $\ln p - \ln q = \ln \frac{p}{q}$.

Check the antiderivative: Take a derivative of $\frac{1}{2}\ln u = \frac{1}{2}\ln(t^2 + 9)$ with respect to t.

Apply the chain rule (Example 1) with $f = \frac{1}{2}\ln u = \frac{1}{2}\ln(t^2 + 9)$ and $u = t^2 + 9$.

$$\frac{df}{dt} = \frac{df}{du}\frac{du}{dt} = \frac{d}{du}\left(\frac{1}{2}\ln u\right)\frac{d}{dt}(t^2 + 9) = \left(\frac{1}{2u}\right)(2t) = \frac{t}{u} = \frac{t}{t^2 + 9}$$

Chapter 2 Problems

Directions: Use a suitable substitution to perform each integral.

❶ $\displaystyle\int \frac{x}{(3x^2 - 5)^3}\,dx$

❷ $\displaystyle\int \sqrt{9t + 4}\,dt$

❸ $\displaystyle\int_{y=1}^{3} (2y - 1)^6\,dy$

❹ $\displaystyle\int_{x=0}^{1} x^3(x^4 + 2)^5\,dx$

❺ $\displaystyle\int \frac{t^2}{7t^3 - 4}\,dt$

❻ $\displaystyle\int \cos(3\theta)\,d\theta$

❼ $\displaystyle\int_{x=0}^{2} \frac{x}{9 - x^2}\,dx$

❽ $\displaystyle\int_{t=2}^{3} \frac{dt}{\sqrt{5t - 6}}$

⑨ $\displaystyle\int we^{w^2}\,dw$

⑩ $\displaystyle\int (3x^2+2)\cos(x^3+2x)\,dx$

⑪ $\displaystyle\int_{\theta=0}^{\sqrt[3]{\pi/4}} \theta^2 \sin(\theta^3)\,d\theta$

⑫ $\displaystyle\int_{\varphi=2/\pi}^{6/\pi} \frac{1}{\varphi^2}\sin\left(\frac{1}{\varphi}\right)d\varphi$

⑬ $\int \sqrt{\varphi} \sin(\varphi\sqrt{\varphi}) \, d\varphi$

⑭ $\int \dfrac{(\sqrt{t}-1)^4}{\sqrt{t}} \, dt$

⑮ $\int_{z=0}^{2} (9z^2 + 12z + 4)^4 \, dz$

⑯ $\int_{\theta=0}^{\pi/6} \cos(8\theta - \pi) \, d\theta$

⑰ $\int \sqrt{\dfrac{x-1}{x^5}}\, dx$

⑱ $\int \dfrac{x^3 + 3x}{(x^4 + 6x^2)^2}\, dx$

⑲ $\int \dfrac{y^5}{\sqrt{y^2 - 5}}\, dy$

⑳ $\int w^5 \sqrt{w^3 + 3}\, dw$

3 Trigonometric Substitutions

In this chapter, we will use the method of substitution that we learned in the previous chapter, but this time every substitution will involve trig functions. There will be two kinds of problems in this chapter. The easier integrals simply involve introducing a suitable new variable via a substitution, much like we did in Chapter 2. Other integrals also involve trig identities. We will tabulate a variety of helpful trig identities, and we will also highlight two sets of identities that are especially useful and worth knowing.

We'll begin by presenting the antiderivatives of the basic trig functions. Recall that in Chapter 1, we only gave the antiderivatives of sine and cosine. We saved the others until now because we can find them using the method of substitution. For example, since the cotangent function is defined by $\cot x = \frac{\cos x}{\sin x}$, the integral of cotangent can be expressed as $\int \cot x \, dx = \int \frac{\cos x}{\sin x} dx$. Since $\frac{d}{dx} \sin x = \cos x$, this suggests making the substitution $u = \sin x$, for which $\frac{du}{dx} = \cos x$. With this substitution, $\int \frac{\cos x}{\sin x} dx = \int \frac{du}{u}$, as we'll explore in Example 1. The derivative of secant is not as straightforward, as shown in Example 2. The antiderivatives of the basic trig functions are listed below.

$$\int \cos x \, dx = \sin x + c \quad , \quad \int \sin x \, dx = -\cos x + c$$

$$\int \tan x \, dx = -\ln|\cos x| + c = \ln|\sec x| + c \quad , \quad \int \cot x \, dx = \ln|\sin x| + c$$

$$\int \sec x \, dx = \ln|\sec x + \tan x| + c \quad , \quad \int \csc x \, dx = \ln|\csc x - \cot x| + c = \ln\left|\tan\frac{x}{2}\right| + c$$

Since we will be looking for derivatives of trig functions in the integrands (in order to rewrite dx in terms of du, as we did in Chapter 2), it will be helpful to be fluent in the derivatives of the basic trig functions. It is worth memorizing the following:

$$\frac{d}{dx} \sin x = \cos x \quad , \quad \frac{d}{dx} \cos x = -\sin x$$

$$\frac{d}{dx} \tan x = \sec^2 x = \frac{1}{\cos^2 x} \quad , \quad \frac{d}{dx} \cot x = -\csc^2 x = -\frac{1}{\sin^2 x}$$

$$\frac{d}{dx} \sec x = \sec x \tan x = \frac{\sin x}{\cos^2 x} \quad , \quad \frac{d}{dx} \csc x = -\csc x \cot x = -\frac{\cos x}{\sin^2 x}$$

Chapter 3 – Trigonometric Substitutions

For example, consider the integral $\int \frac{\sqrt{\tan x}}{\cos^2 x} dx$. Since the secant function is defined as the reciprocal of the cosine function, that is $\sec x = \frac{1}{\cos x}$, this integral is equivalent to $\int \sec^2 x \sqrt{\tan x}\, dx$. Since $\frac{d}{dx} \tan x = \sec^2 x$, this suggests making the substitution $u = \tan x$, for which $\frac{du}{dx} = \sec^2 x$. With this substitution, $\int \sec^2 x \sqrt{\tan x}\, dx = \int \sqrt{u}\, du$, as we'll see in Example 3.

Trig identities are helpful with many integrals involving trig substitutions. A variety of common trig identities are tabulated below.[2]

$$\tan x = \frac{\sin x}{\cos x} \quad , \quad \cot x = \frac{\cos x}{\sin x} \quad , \quad \sec x = \frac{1}{\cos x} \quad , \quad \csc x = \frac{1}{\sin x}$$

$$\sin^2 x + \cos^2 x = 1 \quad , \quad 1 + \tan^2 x = \sec^2 x \quad , \quad 1 + \cot^2 x = \csc^2 x$$

$$\cos(x+y) = \cos x \cos y - \sin x \sin y$$
$$\sin(x+y) = \sin x \cos y + \sin y \cos x \quad , \quad \tan(x+y) = \frac{\tan x + \tan y}{1 - \tan x \tan y}$$

$$\sin(2x) = 2 \sin x \cos x \quad , \quad \cos(2x) = \cos^2 x - \sin^2 x = 2\cos^2 x - 1 = 1 - 2\sin^2 x$$

$$\tan(2x) = \frac{2 \tan x}{1 - \tan^2 x} \quad , \quad \sec(2x) = \frac{\sec^2 x}{2 - \sec^2 x} \quad , \quad \csc(2x) = \frac{\sec x \csc x}{2}$$

$$\sin^2 x = \frac{1 - \cos(2x)}{2} \quad , \quad \cos^2 x = \frac{1 + \cos(2x)}{2} \quad , \quad \tan^2 x = \frac{1 - \cos(2x)}{1 + \cos(2x)}$$

$$\tan\left(\frac{x}{2}\right) = \csc x - \cot x = \frac{\sin x}{1 + \cos x} = \frac{1 - \cos x}{\sin x} = \frac{\tan x}{1 + \sec x}$$

$$\sin(3x) = 3 \sin x - 4 \sin^3 x \quad , \quad \cos(3x) = 4 \cos^3 x - 3 \cos x$$

Notation: $\sin^2 x$ means to square $\sin x$ (that is, the entire function is squared), whereas $\sin(x^2)$ means to square only x (before applying the sine function).

We will use several identities in this book. The more trig identities that you become familiar with, the more tools you'll have at your disposal when performing integrals. The $\sin(2x)$ identity, the Pythagorean identities (like $1 + \tan^2 x = \sec^2 x$), and the $\sin^2 x$ and $\cos^2 x$ identities are especially useful in integration.

---

[2] For a more comprehensive review of trig identities, see Chris McMullen's Trig Identities Practice Workbook with Answers.

For example, consider the integral $\int \sin x \cos x\, dx$. There are two different ways to perform this integral. One way is to use the double-angle sine identity $\sin(2x) = 2\sin x \cos x$. Another way is to let $u = \sin x$ such that $\frac{du}{dx} = \cos x$. See Example 4.

The identities $\sin^2 x = \frac{1-\cos(2x)}{2}$ and $\cos^2 x = \frac{1+\cos(2x)}{2}$ can be especially helpful when you see even powers of sine or cosine, like $\int \sin^2 x\, dx$ or like $\int \cos^4 x\, dx$, as shown in Examples 5-6. However, when there are odd powers, a different substitution may be simpler, as is the case with $\int \sin^3 x\, dx$ and $\int \cos^4 x \sin x\, dx$. See Examples 7-8.

The Pythagorean identities $\sin^2 x + \cos^2 x = 1$, $1 + \tan^2 x = \sec^2 x$, and $1 + \cot^2 x = \csc^2 x$ are among the most useful identities. These are really all the same identity. The identity $\sin^2 x + \cos^2 x = 1$ represents the Pythagorean theorem for the right triangle typically drawn on a unit circle. If you divide both sides by $\cos^2 x$, you get $\tan^2 x + 1 = \sec^2 x$. If instead you divide both sides by $\sin^2 x$, you get $1 + \cot^2 x = \csc^2 x$. These identities can be used two ways. The most obvious way to use these identities is to trade one trig function for another. For example, the integral $\int \sin^3 x\, dx$ is easier after rewriting it as $\int \sin^3 x\, dx = \int \sin^2 x \sin x\, dx = \int (1 - \cos^2 x)\sin x\, dx$, as we'll see in Example 7. A less obvious, but very important, way to use the Pythagorean identities is when the integrand involves expressions like $4 - x^2$, $x^2 + 25$, or $x^2 - 9$, as we'll see in Examples 9-11. For example, $\int \frac{dx}{\sqrt{9-x^2}}$ can be performed using $\sin^2 x + \cos^2 x = 1$, as we'll see in Example 9. Note that this integrand doesn't even involve a trig function, yet it can be integrated by using a trig identity.

Yet another way that a trig identity may help is with the answer to an indefinite integral. **Two different answers may appear to be considerably different, yet actually be equivalent.** (This is true for **indefinite** integrals, but not for definite integrals; the answer to a definite integral is a number.) For example, suppose that online integral software tells you that the answer to an integral is $\tan^2 x + c$, but the book's answer to the same integral is $\sec^2 x + c$. Although the tangent function is quite different from the secant function, these two answers are **equivalent**. How? Because $1 + \tan^2 x = \sec^2 x$. This identity means that $\sec^2 x + c = \tan^2 x + 1 + c$, and 1 is just a constant.

Chapter 3 – Trigonometric Substitutions

The two constants (both labeled with the same letter c) can't be the same value in these two answers. If we use subscripts to write $\tan^2 x + c_1$ and $\sec^2 x + c_2$, then the latter expression becomes $\sec^2 x + c_2 = \tan^2 x + 1 + c_2$. By comparison, we see that $c_1 = 1 + c_2$. But this doesn't really matter. It's just a general constant. Whether we say that the answer to the integral is $\tan^2 x + $ a constant or $\sec^2 x + $ a constant doesn't really matter. The answers are equivalent. (But if we evaluate a definite integral, then the constant will subtract out and we will get the same numerical value either way.)

Tips for making effective trig substitutions:
- Remember to identify the **derivative** because you need to replace dx with du. For example, in $\int \frac{\cos x}{\sin x} dx$, we may let $u = \sin x$ and then the numerator will include the derivative since $\frac{du}{dx} = \cos x$. See Example 1.
- Often, when you see **powers** of trig functions, as in $\int \cos^4 x \, dx$ or $\int \sin^3 x \, dx$, you want to use an identity to reduce the power. For example, for $\int \cos^4 x \, dx$, use $\cos^2 x = \frac{1+\cos(2x)}{2}$ (Example 6), and for $\int \sin^3 x \, dx$, use $\sin^2 x + \cos^2 x = 1$ (Example 7). An exception is an integral of the form $\int \cos^4 x \sin x \, dx$, where $-\sin x$ is the derivative of $\cos x$. This integral can be done using $u = \cos x$. See Example 8.
- **The arguments of a trig identity can be changed**, provided that every argument is changed the same way. For example, $\sin^2 x = \frac{1-\cos(2x)}{2}$ and $\sin^2 \left(\frac{x}{2}\right) = \frac{1-\cos x}{2}$ are both valid. In the latter case, we divided every argument by two. Similarly, $\sin(3x) = 3 \sin x - 4 \sin^3 x$ may be adapted to $\sin x = 3 \sin \left(\frac{x}{3}\right) - 4 \sin^3 \left(\frac{x}{3}\right)$.
- Don't be afraid to separate one integral into **multiple integrals**. For example, when we use $\sin^2 x + \cos^2 x = 1$ to write $\int \sin^3 x \, dx = \int \sin^2 x \sin x \, dx = \int (1 - \cos^2 x) \sin x \, dx = \int \sin x \, dx - \int \cos^2 x \sin x \, dx$, the new integrals are simpler than the original integral (Example 7).
- If you see $a^2 - x^2$ ($a > x$), let $x = a \sin \theta$ and $dx = a \cos \theta \, d\theta$. If you see $x^2 + a^2$, let $x = a \tan \theta$ and $dx = a \sec^2 \theta \, d\theta$. If you see $x^2 - a^2$ ($x > a$), let $x = a \sec \theta$ and $dx = a \sec \theta \tan \theta \, d\theta$. The idea is that one of the **Pythagorean identities** will condense two terms down to a single term. See Examples 9-11.

Techniques of Integration Calculus Practice Workbook

Example 1. Find the solution to the indefinite integral below.
$$\int \cot\theta \, d\theta$$
Solution: Use the definition of cotangent: $\cot\theta = \frac{\cos\theta}{\sin\theta}$.
$$\int \cot\theta \, d\theta = \int \frac{\cos\theta}{\sin\theta} d\theta$$
Since cosine is the derivative of sine, let $u = \sin\theta$, such that $du = \cos\theta \, d\theta$.
$$\int \frac{\cos\theta}{\sin\theta} d\theta = \int \frac{du}{u} = \ln|u| + c = \ln|\sin\theta| + c$$
Check the answer: Take a derivative of the answer with respect to θ and verify that this matches the original integrand. Apply the chain rule (Chapter 1, Examples 3 and 7) with $f = \ln u + c = \ln\sin\theta + c$ and $u = \sin\theta$.
$$\frac{df}{d\theta} = \frac{df}{du}\frac{du}{d\theta} = \frac{d}{du}(\ln u + c)\frac{d}{d\theta}\sin\theta = \frac{\cos\theta}{u} = \frac{\cos\theta}{\sin\theta} = \cot\theta$$

Example 2. Perform the definite integral below.
$$\int_{\varphi=0}^{\pi/4} \sec\varphi \, d\varphi$$
Notation: θ and φ are the lowercase Greek letters theta and phi.
Solution: Although the derivative of secant is straightforward and easy, its integral serves as a reminder that many seemingly simple integrals are nonobvious (or even algebraically impossible, like $\int e^{-x^2} dx$). You would have to be very clever to figure this out yourself. It turns out that multiplying the numerator and denominator each by $\sec\varphi + \tan\varphi$ leads to a substitution that works.[3]
$$\int_{\varphi=0}^{\pi/4} \sec\varphi \, d\varphi = \int_{\varphi=0}^{\pi/4} \sec\varphi \frac{\sec\varphi + \tan\varphi}{\sec\varphi + \tan\varphi} d\varphi = \int_{\varphi=0}^{\pi/4} \frac{\sec^2\varphi + \sec\varphi \tan\varphi}{\sec\varphi + \tan\varphi} d\varphi$$

---

[3] One way to look at this is that $\int \sec x \, dx = \int \frac{dx}{\cos x}$ involves a reciprocal trig function, so we would expect the antiderivative to involve a logarithm similar to Example 1. So we want to make a substitution such that the integral acquires the form $\int \frac{du}{u}$. It wouldn't be easy for most people to figure out that $u = \sec x + \tan x$ accomplishes the task, so if this seems nonobvious to you, that's because it actually is.

Chapter 3 – Trigonometric Substitutions

Now the numerator is the derivative of the denominator. Let $u = \sec\varphi + \tan\varphi$, such that $du = (\sec\varphi \tan\varphi + \sec^2\varphi)d\varphi$. (Just as in the previous chapter, we're simply taking the derivative $\frac{du}{d\varphi}$ to find the equation that relates du to $d\varphi$. If you're struggling to see how we found du, try taking a derivative of u with respect to φ, and if you still need more help with this, review the examples from the previous chapter.) Plug the given limits (0 and $\frac{\pi}{4}$) into the equation $u = \sec\varphi + \tan\varphi$ to find the new limits of integration: $\sec 0 + \tan 0 = 1 + 0 = 1$ and $\sec\frac{\pi}{4} + \tan\frac{\pi}{4} = \sqrt{2} + 1 = 1 + \sqrt{2}$.

$$\int_{\varphi=0}^{\pi/4} \frac{\sec^2\varphi + \sec\varphi \tan\varphi}{\sec\varphi + \tan\varphi} d\varphi = \int_{\varphi=0}^{\pi/4} \frac{\sec\varphi \tan\varphi + \sec^2\varphi}{\sec\varphi + \tan\varphi} d\varphi = \int_{u=1}^{1+\sqrt{2}} \frac{du}{u} = [\ln|u|]_{u=1}^{1+\sqrt{2}}$$

$$= \ln(1+\sqrt{2}) - \ln 1 = \ln(1+\sqrt{2}) - 0 = \ln(1+\sqrt{2}) \approx 0.881$$

Alternate answer: $\frac{\ln(2\sqrt{2}+3)}{2} \approx 0.881$. This works because $\frac{\ln(2\sqrt{2}+3)}{2} = \ln\sqrt{2\sqrt{2}+3}$ and because $\sqrt{2\sqrt{2}+3} = 1 + \sqrt{2}$; you can verify that $(1+\sqrt{2})^2 = 2\sqrt{2} + 3$.

Note: It's worth memorizing that $\int \sec\varphi\, d\varphi = \int \frac{d\varphi}{\cos\varphi} = \ln|\sec\varphi + \tan\varphi| + c$ because this integral is difficult to work out from scratch. (But see Chapter 4, Example 5.) Check the antiderivative: Take a derivative of $\ln(\sec\varphi + \tan\varphi)$ with respect to φ. Apply the chain rule (Example 1) with $f = \ln u$ and $u = \sec\varphi + \tan\varphi$.

$$\frac{df}{d\varphi} = \frac{df}{du}\frac{du}{d\varphi} = \frac{d}{du}\ln u \frac{d}{d\varphi}(\sec\varphi + \tan\varphi) = \frac{1}{u}(\sec\varphi \tan\varphi + \sec^2\varphi)$$

$$\frac{\sec\varphi \tan\varphi + \sec^2\varphi}{\sec\varphi + \tan\varphi} = \frac{\sec\varphi(\sec\varphi + \tan\varphi)}{\sec\varphi + \tan\varphi} = \sec\varphi = \frac{1}{\cos\varphi}$$

Example 3. Find the solution to the indefinite integral below.

$$\int \frac{\sqrt{\tan x}}{\cos^2 x} dx$$

Solution: Use the definition of secant: $\sec x = \frac{1}{\cos x}$.

$$\int \frac{\sqrt{\tan x}}{\cos^2 x} dx = \int \sec^2 x \sqrt{\tan x}\, dx$$

Since $\sec^2 x$ is the derivative of $\tan x$, let $u = \tan x$, such that $du = \sec^2 x\, dx$.

$$\int \sec^2 x \sqrt{\tan x}\, dx = \int \sqrt{u}\, du = \int u^{1/2}\, du = \frac{u^{\frac{1}{2}+1}}{\frac{1}{2}+1} + c$$

$$= \frac{u^{3/2}}{3/2} + c = \frac{2}{3}u^{3/2} + c = \frac{2}{3}\tan^{3/2} x + c = \frac{2}{3}\tan x \sqrt{\tan x} + c$$

Note: Recall from algebra that $u^{3/2} = u^1 u^{1/2} = u\sqrt{u}$.

Check the answer: Take a derivative of the answer with respect to x and verify that this matches the original integrand. Apply the chain rule (Example 1) with $f = \frac{2}{3}u^{3/2} + c = \frac{2}{3}\tan^{3/2} x + c$ and $u = \tan x$.

$$\frac{df}{dx} = \frac{df}{du}\frac{du}{dx} = \frac{d}{du}\left(\frac{2}{3}u^{3/2} + c\right)\frac{d}{dx}\tan x = u^{1/2}\sec^2 x = \sec^2 x \sqrt{\tan x} = \frac{\sqrt{\tan x}}{\cos^2 x}$$

Example 4. Perform the definite integral below.

$$\int_{\theta=0}^{\pi/6} \sin\theta \cos\theta \, d\theta$$

Solution: We will do this integral using two different methods. First, we will use the identity $\sin(2\theta) = 2\sin\theta\cos\theta$.

$$\int_{\theta=0}^{\pi/6} \sin\theta \cos\theta \, d\theta = \int_{\theta=0}^{\pi/6} \frac{\sin(2\theta)}{2} d\theta = \frac{1}{2}\int_{\theta=0}^{\pi/6} \sin(2\theta) \, d\theta$$

Let $u = 2\theta$, such that $du = 2d\theta$, for which $\frac{du}{2} = d\theta$. Plug the given limits (0 and $\frac{\pi}{6}$) into the equation $u = 2\theta$ to find the new limits of integration: $2(0) = 0$ and $2\left(\frac{\pi}{6}\right) = \frac{\pi}{3}$. Replace 2θ with u and also replace $d\theta$ with $\frac{du}{2}$.

$$\frac{1}{2}\int_{\theta=0}^{\pi/6} \sin(2\theta) \, d\theta = \frac{1}{2}\int_{u=0}^{\pi/3} \frac{\sin u}{2} du = \frac{1}{4}\int_{u=0}^{\pi/3} \sin u \, du = \frac{1}{4}[-\cos u]_{u=0}^{\pi/3}$$

$$= -\frac{1}{4}\cos\frac{\pi}{3} - \left(-\frac{1}{4}\right)\cos 0 = -\frac{1}{4}\left(\frac{1}{2}\right) + \frac{1}{4}(1) = -\frac{1}{8} + \frac{1}{4} = -\frac{1}{8} + \frac{2}{8} = \frac{1}{8} = 0.125$$

Alternate solution: Let $u = \sin\theta$, such that $du = \cos\theta \, d\theta$. Plug the given limits (0 and $\frac{\pi}{6}$) into the equation $u = \sin\theta$ to find the new limits: $\sin 0 = 0$ and $\sin\frac{\pi}{6} = \frac{1}{2}$.

$$\int_{\theta=0}^{\pi/6} \sin\theta \cos\theta \, d\theta = \int_{u=0}^{1/2} u \, du = \left[\frac{u^2}{2}\right]_{u=0}^{1/2} = \frac{1}{2}\left(\frac{1}{2}\right)^2 - \frac{1}{2}(0)^2 = \frac{1}{2}\left(\frac{1}{4}\right) - 0 = \frac{1}{8} = 0.125$$

Check the antiderivative: Take a derivative of $\frac{\sin^2\theta}{2}$ with respect to θ. Apply the chain rule (Example 1) with $f = \frac{u^2}{2} = \frac{\sin^2\theta}{2}$ and $u = \sin\theta$.

$$\frac{df}{d\theta} = \frac{df}{du}\frac{du}{d\theta} = \frac{d}{du}\frac{u^2}{2}\frac{d}{d\theta}\sin\theta = u\cos\theta = \sin\theta\cos\theta$$

Example 5. Find the solution to the indefinite integral below.

$$\int \sin^2\varphi\, d\varphi$$

Solution: Use the identity $\sin^2\varphi = \frac{1-\cos(2\varphi)}{2}$.

$$\int \sin^2\varphi\, d\varphi = \int \frac{1-\cos(2\varphi)}{2}\, d\varphi = \frac{1}{2}\int [1-\cos(2\varphi)]\, d\varphi = \frac{1}{2}\int d\varphi - \frac{1}{2}\int \cos(2\varphi)\, d\varphi$$

In the second integral, let $u = 2\varphi$, such that $du = 2d\varphi$, for which $\frac{du}{2} = d\varphi$. Replace 2φ with u and also replace $d\varphi$ with $\frac{du}{2}$.

$$\frac{1}{2}\int d\varphi - \frac{1}{2}\int \cos(2\varphi)\, d\varphi = \frac{\varphi}{2} - \frac{1}{2}\int \frac{\cos u}{2}\, du = \frac{\varphi}{2} - \frac{1}{4}\int \cos u\, du$$

$$= \frac{\varphi}{2} - \frac{1}{4}\sin u + c = \frac{\varphi}{2} - \frac{1}{4}\sin(2\varphi) + c = \frac{\varphi}{2} - \frac{1}{2}\sin\varphi\cos\varphi + c$$

Note: We used the identity $\sin(2\varphi) = 2\sin\varphi\cos\varphi$.

Check the answer: Take a derivative of the answer with respect to φ and verify that this matches the original integrand.

$$\frac{d}{d\varphi}\left[\frac{\varphi}{2} - \frac{1}{4}\sin(2\varphi) + c\right] = \frac{1}{2} - \frac{1}{4}(2)\cos(2\varphi) = \frac{1}{2} - \frac{1}{2}\cos(2\varphi)$$

$$= \frac{1}{2} - \frac{1}{2}(1 - 2\sin^2\varphi) = \frac{1}{2} - \frac{1}{2} + \sin^2\varphi = \sin^2\varphi$$

Note: We used the identity $\cos(2\varphi) = 1 - 2\sin^2\varphi$.

Example 6. Perform the definite integral below.

$$\int_{t=\pi/3}^{\pi} \cos^4 t\, dt$$

Solution: With only even powers of sine and cosine, try using the $\sin^2 t = \frac{1-\cos(2t)}{2}$ and $\cos^2 t = \frac{1+\cos(2t)}{2}$ formulas.

$$\int_{t=\pi/3}^{\pi} \cos^4 t \, dt = \int_{t=\pi/3}^{\pi} \cos^2 t \cos^2 t \, dt = \int_{t=\pi/3}^{\pi} \left[\frac{1+\cos(2t)}{2}\right]\left[\frac{1+\cos(2t)}{2}\right] dt$$

$$= \frac{1}{4} \int_{t=\pi/3}^{\pi} [1+\cos(2t)][1+\cos(2t)] \, dt = \frac{1}{4} \int_{t=\pi/3}^{\pi} [1 + 2\cos(2t) + \cos^2(2t)] \, dt$$

$$= \frac{1}{4} \int_{t=\pi/3}^{\pi} dt + \frac{1}{2} \int_{t=\pi/3}^{\pi} \cos(2t) \, dt + \frac{1}{4} \int_{t=\pi/3}^{\pi} \cos^2(2t) \, dt$$

For the middle integral, let $u = 2t$ such that $du = 2dt$ or $\frac{du}{2} = dt$, with the new limits from $u = 2\left(\frac{\pi}{3}\right) = \frac{2\pi}{3}$ to $2(\pi) = 2\pi$, and in the right integral, use the cosine squared identity again: $\cos^2(2t) = \frac{1+\cos(4t)}{2}$.

$$\int_{t=\pi/3}^{\pi} \cos^4 t \, dt = \frac{1}{4} \int_{t=\pi/3}^{\pi} dt + \frac{1}{2} \int_{u=2\pi/3}^{2\pi} \frac{\cos u}{2} \, du + \frac{1}{4} \int_{t=\pi/3}^{\pi} \frac{1+\cos(4t)}{2} \, dt$$

$$= \frac{1}{4}[t]_{t=\pi/3}^{\pi} + \frac{1}{4} \int_{u=2\pi/3}^{2\pi} \cos u \, du + \frac{1}{8} \int_{t=\pi/3}^{\pi} dt + \frac{1}{8} \int_{t=\pi/3}^{\pi} \cos(4t) \, dt$$

In the right integral, let $w = 4t$ such that $dw = 4dt$ or $\frac{dw}{4} = dt$, with the new limits from $w = 4\left(\frac{\pi}{3}\right) = \frac{4\pi}{3}$ to $4(\pi) = 4\pi$.

$$\int_{t=\pi/3}^{\pi} \cos^4 t \, dt = \frac{1}{4}\left(\pi - \frac{\pi}{3}\right) + \frac{1}{4}[\sin u]_{u=2\pi/3}^{2\pi} + \frac{1}{8}[t]_{t=\pi/3}^{\pi} + \frac{1}{8} \int_{w=4\pi/3}^{4\pi} \frac{\cos w}{4} \, dw$$

$$= \frac{\pi}{4} - \frac{\pi}{12} + \frac{\sin(2\pi)}{4} - \frac{1}{4}\sin\left(\frac{2\pi}{3}\right) + \frac{1}{8}\left(\pi - \frac{\pi}{3}\right) + \frac{1}{32} \int_{u=4\pi/3}^{4\pi} \cos w \, dw$$

$$= \frac{3\pi}{12} - \frac{\pi}{12} + \frac{0}{4} - \frac{1}{4}\left(\frac{\sqrt{3}}{2}\right) + \frac{\pi}{8} - \frac{\pi}{24} + \frac{1}{32}[\sin w]_{w=4\pi/3}^{4\pi}$$

$$= \frac{2\pi}{12} + 0 - \frac{\sqrt{3}}{8} + \frac{3\pi}{24} - \frac{\pi}{24} + \frac{\sin(4\pi)}{32} - \frac{1}{32}\sin\left(\frac{4\pi}{3}\right)$$

$$= \frac{4\pi}{24} - \frac{\sqrt{3}}{8} + \frac{2\pi}{24} + \frac{0}{32} - \frac{1}{32}\left(-\frac{\sqrt{3}}{2}\right) = \frac{6\pi}{24} - \frac{\sqrt{3}}{8} + 0 + \frac{\sqrt{3}}{64}$$

Chapter 3 – Trigonometric Substitutions

$$= \frac{\pi}{4} - \frac{8\sqrt{3}}{64} + \frac{\sqrt{3}}{64} = \frac{\pi}{4} - \frac{7\sqrt{3}}{64} \approx 0.596$$

Check the antiderivative: There were four integrals in all. Find the antiderivative from each (that is, before numbers were plugged in) to get the complete antiderivative:

$$f = \frac{t}{4} + \frac{\sin u}{4} + \frac{t}{8} + \frac{\sin w}{32} = \frac{2t}{8} + \frac{t}{8} + \frac{\sin u}{4} + \frac{\sin w}{32} = \frac{3t}{8} + \frac{\sin u}{4} + \frac{\sin w}{32} = \frac{3t}{8} + \frac{\sin(2t)}{4} + \frac{\sin(4t)}{32}$$

with $u = 2t$ and $w = 4t$. Take a derivative with respect to t. Apply the chain rule (Example 1) to each of the last two terms.

$$\frac{df}{dt} = \frac{d}{dt}\left[\frac{3t}{8} + \frac{\sin(2t)}{4} + \frac{\sin(4t)}{32}\right] = \frac{3}{8} + \frac{\cos(2t)}{2} + \frac{\cos(4t)}{8}$$

Use $\cos(2t) = 2\cos^2 t - 1$, which can also be expressed as $\cos(4t) = 2\cos^2(2t) - 1$.

$$= \frac{3}{8} + \frac{1}{2}(2\cos^2 t - 1) + \frac{1}{8}[2\cos^2(2t) - 1]$$

$$= \frac{3}{8} + \cos^2 t - \frac{1}{2} + \frac{1}{4}\cos^2(2t) - \frac{1}{8}$$

$$= \frac{3}{8} - \frac{1}{2} - \frac{1}{8} + \cos^2 t + \frac{1}{4}\cos^2(2t)$$

Use $\cos(2t) = 2\cos^2 t - 1$ again. Here we have $\cos^2(2t) = (2\cos^2 t - 1)^2$.

$$= \frac{3}{8} - \frac{4}{8} - \frac{1}{8} + \cos^2 t + \frac{1}{4}(2\cos^2 t - 1)^2$$

$$= \frac{3 - 4 - 1}{8} + \cos^2 t + \frac{1}{4}(4\cos^4 t - 4\cos^2 t + 1)$$

$$= -\frac{2}{8} + \cos^2 t + \cos^4 t - \cos^2 t + \frac{1}{4} = -\frac{1}{4} + \cos^4 t + \frac{1}{4} = \cos^4 t$$

Example 7. Find the solution to the indefinite integral below.

$$\int \sin^3 \theta \, d\theta$$

Solution: In contrast to Example 6, here the exponent is odd, so we'll adopt a different strategy. Use the identity $\sin^2 \theta + \cos^2 \theta = 1$, which may be expressed as $\sin^2 \theta = 1 - \cos^2 \theta$.

$$\int \sin^3 \theta \, d\theta = \int \sin^2 \theta \sin \theta \, d\theta = \int (1 - \cos^2 \theta) \sin \theta \, d\theta$$

$$= \int \sin \theta \, d\theta - \int \cos^2 \theta \sin \theta \, d\theta$$

In the right integral, let $u = \cos\theta$, such that $du = -\sin\theta\, d\theta$. The minus signs cancel.

$$\int \sin\theta\, d\theta - \int \cos^2\theta \sin\theta\, d\theta = -\cos\theta + \int u^2\, du$$

$$= -\cos\theta + \frac{u^3}{3} + c = -\cos\theta + \frac{\cos^3\theta}{3} + c$$

Note: Using various trig identities, the answers to indefinite integrals may appear in a variety of forms. For example, we could use the identity $\cos(3\theta) = 4\cos^3\theta - 3\cos\theta$ to express the answer in terms of $\cos\theta$ and $\cos(3\theta)$ (instead of $\cos\theta$ and $\cos^3\theta$). Check the answer: Take a derivative of the answer with respect to θ and verify that this matches the original integrand. We applied the chain rule (Example 1) to the first term below: $\frac{d}{d\theta}\cos^3\theta = \frac{d}{du}u^3 \frac{d}{d\theta}\cos\theta = (3u^2)(-\sin\theta) = -3\cos^2\theta \sin\theta$.

$$\frac{d}{d\theta}\left(-\cos\theta + \frac{\cos^3\theta}{3} + c\right) = \sin\theta + \cos^2\theta\,(-\sin\theta) = \sin\theta - \sin\theta\cos^2\theta$$

$$= \sin\theta\,(1 - \cos^2\theta) = \sin\theta \sin^2\theta = \sin^3\theta$$

Example 8. Perform the definite integral below.

$$\int_{y=\pi/3}^{\pi} \cos^4 y \sin y\, dy$$

Solution: In contrast to Example 6, here the power of sine is odd (since it is 'raised' to the first power), so we won't use the double-angle formulas. Since $-\sin y$ is the derivative of $\cos y$, this suggests letting $u = \cos y$ such that $du = -\sin y$. Plug the given limits ($\frac{\pi}{3}$ and π) into the equation $u = \cos y$ to find the new limits of integration: $\cos\frac{\pi}{3} = \frac{1}{2}$ and $\cos\pi = -1$.

$$\int_{y=\pi/3}^{\pi} \cos^4 y \sin y\, dy = \int_{u=1/2}^{-1} u^4\,(-du) = -\int_{u=1/2}^{-1} u^4\, du = -\left[\frac{u^5}{5}\right]_{u=1/2}^{-1}$$

$$= -\left[\frac{(-1)^5}{5} - \frac{1}{5}\left(\frac{1}{2}\right)^5\right] = -\left[-\frac{1}{5} - \frac{1}{5}\left(\frac{1}{32}\right)\right] = \frac{1}{5} + \frac{1}{160} = \frac{32}{160} + \frac{1}{160} = \frac{33}{160} \approx 0.206$$

Check the antiderivative: Take a derivative of $-\frac{\cos^5 y}{5}$ with respect to y. Apply the chain rule (Example 1) with $f = -\frac{u^5}{5} = -\frac{\cos^5 y}{5}$ and $u = \cos y$.

Chapter 3 – Trigonometric Substitutions

$$\frac{df}{dy} = \frac{df}{du}\frac{du}{dy} = \frac{d}{du}\left(-\frac{u^5}{5}\right)\frac{d}{dy}\cos y = (-u^4)(-\sin y) = \cos^4 y \sin y$$

Example 9. Find the solution to the indefinite integral below (where $|x| < 3$).[4]

$$\int \frac{dx}{\sqrt{9-x^2}}$$

Solution: According to the last bullet point on page 36, if we see an expression of the form $a^2 - x^2$, we should try[5] $x = a \sin\theta$ and $dx = a \cos\theta \, d\theta$. First rewrite $9 - x^2$ as $3^2 - x^2$ and then compare it with $a^2 - x^2$ to see that $a = 3$. Let $x = 3\sin\theta$ such that $dx = 3\cos\theta \, d\theta$.

$$\int \frac{dx}{\sqrt{9-x^2}} = \int \frac{3\cos\theta \, d\theta}{\sqrt{9-(3\sin\theta)^2}} = \int \frac{3\cos\theta \, d\theta}{\sqrt{9-9\sin^2\theta}} = \int \frac{3\cos\theta \, d\theta}{\sqrt{9(1-\sin^2\theta)}}$$

$$= \int \frac{3\cos\theta \, d\theta}{3\sqrt{1-\sin^2\theta}} = \int \frac{\cos\theta \, d\theta}{\sqrt{\cos^2\theta}} = \int \frac{\cos\theta \, d\theta}{\cos\theta} = \int d\theta = \theta + c = \sin^{-1}\left(\frac{x}{3}\right) + c$$

Note: We used the equation $x = 3\sin\theta$ to determine that $\sin^{-1}\left(\frac{x}{3}\right) = \theta$ in the last step. (First we divided both sides by 3 to get $\frac{x}{3} = \sin\theta$, and then we took the inverse sine of both sides.)

Check the answer: Take a derivative of the answer with respect to x and verify that this matches the original integrand. Apply the chain rule (Example 1) with $f = \sin^{-1} u + c$ and $u = \frac{x}{3}$. Recall from first-semester calculus that $\frac{d}{du}\sin^{-1} u = \frac{1}{\sqrt{1-u^2}}$ provided that $-\frac{\pi}{2} < \sin^{-1} u < \frac{\pi}{2}$.

---

[4] If you're so fluent with your derivatives that you know $\frac{d}{du}\sin^{-1} u$ off the top of your head, pretend for a minute that you don't know that. This particular integral happens to be trivial if you know what the derivative of the inverse sine is. But the idea of this example is to illustrate the strategy for how to solve a more general problem involving an expression of the form $a^2 - x^2$. This is a practical strategy which arises in a variety of applications (such as electric field integrals in physics), and in most cases the answer to an integral wouldn't be as simple as an inverse trig function.

[5] Why? Since $\sin^2\theta + \cos^2\theta = 1$, it follows that $\cos^2\theta = 1 - \sin^2\theta$. So, for example, if we see the expression $1 - x^2$, if we make the substitution $x = \sin\theta$, we get $1 - x^2 = 1 - \sin^2\theta = \cos^2\theta$, which condenses the two terms $1 - x^2$ down to the single term $\cos^2\theta$. Here, we don't have $1 - x^2$; we have $9 - x^2$, which equates to $3^2 - x^2$. The same idea applies if we work with $3^2 \sin^2\theta + 3^2 \cos^2\theta = 3^2$. So here we let $x = 3\sin\theta$ so that $3^2 - x^2 = 3^2 - 3^2\sin^2\theta = 3^2\cos^2\theta$. Again, the main idea is that the trig identity $\sin^2\theta + \cos^2\theta = 1$ condenses two terms of the form $a^2 - x^2$ down to a single term.

Techniques of Integration Calculus Practice Workbook

$$\frac{df}{dx} = \frac{df}{du}\frac{du}{dx} = \frac{d}{du}\sin^{-1}u\frac{dx}{dx\,3} = \frac{1}{\sqrt{1-u^2}}\left(\frac{1}{3}\right) = \frac{1}{3\sqrt{1-\left(\frac{x}{3}\right)^2}}$$

$$= \frac{1}{\sqrt{9}\sqrt{1-\frac{x^2}{9}}} = \frac{1}{\sqrt{9\left(1-\frac{x^2}{9}\right)}} = \frac{1}{\sqrt{9-x^2}}$$

Example 10. Perform the definite integral below.

$$\int_{y=0}^{2} \frac{dy}{(y^2+4)^{3/2}}$$

Solution: According to the last bullet point on page 36, if we see an expression of the form $y^2 + a^2$, we should try[6] $y = a\tan\theta$ and $dy = a\sec^2\theta\, d\theta$. First rewrite $y^2 + 4$ as $y^2 + 2^2$ and then compare it with $y^2 + a^2$ to see that $a = 2$. Let $y = 2\tan\theta$ such that $dy = 2\sec^2\theta\, d\theta$. Divide by 2 on both sides of $y = 2\tan\theta$ to get $\frac{y}{2} = \tan\theta$ and then take the inverse tangent of both sides: $\tan^{-1}\left(\frac{y}{2}\right) = \theta$. Plug the given limits (0 and 2) into the equation $\theta = \tan^{-1}\left(\frac{y}{2}\right)$ to find the new limits of integration: $\tan^{-1}0 = 0$ and $\tan^{-1}1 = \frac{\pi}{4}$. Note that $(\sec^2\theta)^{3/2} = \sec^3\theta$ because $(p^2)^{3/2} = p^{2(3/2)} = p^3$.

$$\int_{y=0}^{2}\frac{dy}{(y^2+4)^{3/2}} = \int_{\theta=0}^{\pi/4}\frac{2\sec^2\theta}{[(2\tan\theta)^2+4]^{3/2}}d\theta = \int_{\theta=0}^{\pi/4}\frac{2\sec^2\theta}{(4\tan^2\theta+4)^{3/2}}d\theta$$

$$= \int_{\theta=0}^{\pi/4}\frac{2\sec^2\theta}{[4(\tan^2\theta+1)]^{3/2}}d\theta = \int_{\theta=0}^{\pi/4}\frac{2\sec^2\theta}{4^{3/2}(\tan^2\theta+1)^{3/2}}d\theta = \int_{\theta=0}^{\pi/4}\frac{2\sec^2\theta}{8(\sec^2\theta)^{3/2}}d\theta$$

$$= \frac{1}{4}\int_{\theta=0}^{\pi/4}\frac{\sec^2\theta\, d\theta}{\sec^3\theta} = \frac{1}{4}\int_{\theta=0}^{\pi/4}\frac{d\theta}{\sec\theta} = \frac{1}{4}\int_{\theta=0}^{\pi/4}\cos\theta\, d\theta = \frac{1}{4}[\sin\theta]_{\theta=0}^{\pi/4}$$

---

[6] Why? Since $\tan^2\theta + 1 = \sec^2\theta$. So, for example, if we see the expression $y^2 + 1$, if we make the substitution $y = \tan\theta$, we get $y^2 + 1 = \tan^2\theta + 1 = \sec^2\theta$, which condenses the two terms $y^2 + 1$ down to the single term $\sec^2\theta$. Here, we don't have $y^2 + 1$; we have $y^2 + 4$, which equates to $y^2 + 2^2$. The same idea applies if we work with $2^2\tan^2\theta + 2^2\sec^2\theta = 2^2$. So here we let $y = 2\tan\theta$ so that $y^2 + 2^2 = 2^2\tan^2\theta + 2^2 = 2^2\sec^2\theta$. Again, the main idea is that the trig identity $\tan^2\theta + 1 = \sec^2\theta$ condenses two terms of the form $y^2 + a^2$ down to a single term.

$$= \frac{1}{4}\sin\frac{\pi}{4} - \frac{1}{4}\sin 0 = \frac{1}{4}\frac{\sqrt{2}}{2} - \frac{1}{4}(0) = \frac{\sqrt{2}}{8} \approx 0.177$$

Check the antiderivative: We found that the indefinite integral equals $\frac{\sin\theta}{4}$, where θ equals $\theta = \tan^{-1}\left(\frac{y}{2}\right)$. So the indefinite integral equals $\frac{1}{4}\sin\left[\tan^{-1}\left(\frac{y}{2}\right)\right]$. Recall the trig identity[7] $\sin(\tan^{-1} u) = \frac{u}{\sqrt{1+u^2}}$. With this, our indefinite integral equals $\frac{y/2}{4\sqrt{1+\left(\frac{y}{2}\right)^2}} =$

$\frac{y}{8\sqrt{1+\frac{y^2}{4}}} = \frac{y}{\sqrt{64}\sqrt{1+\frac{y^2}{4}}} = \frac{y}{\sqrt{64\left(1+\frac{y^2}{4}\right)}} = \frac{y}{\sqrt{64+16y^2}}$. To check our answer, take a derivative of this indefinite integral with respect to y. Apply the **quotient rule**.

$$\frac{df}{dy} = \frac{d}{dy}\frac{y}{\sqrt{64+16y^2}} = \frac{\sqrt{64+16y^2}\frac{dy}{dy} - y\frac{d}{dy}\sqrt{64+16y^2}}{64+16y^2}$$

$$= \frac{\sqrt{64+16y^2}(1) - y\frac{32y}{2\sqrt{64+16y^2}}}{64+16y^2} = \frac{\sqrt{64+16y^2} - \frac{16y^2}{\sqrt{64+16y^2}}}{64+16y^2}$$

$$= \frac{\sqrt{64+16y^2}\frac{\sqrt{64+16y^2}}{\sqrt{64+16y^2}} - \frac{16y^2}{\sqrt{64+16y^2}}}{64+16y^2} = \frac{\frac{64+16y^2}{\sqrt{64+16y^2}} - \frac{16y^2}{\sqrt{64+16y^2}}}{64+16y^2} = \frac{\frac{64}{\sqrt{64+16y^2}}}{64+16y^2}$$

$$= \frac{64}{\sqrt{64+16y^2}} \div (64+4y^2) = \frac{64}{\sqrt{64+16y^2}}\frac{1}{64+16y^2} = \frac{64}{(64+16y^2)^{3/2}}$$

$$= \frac{64}{[16(4+y^2)]^{3/2}} = \frac{64}{16^{3/2}(4+y^2)^{3/2}} = \frac{64}{64(4+y^2)^{3/2}} = \frac{1}{(4+y^2)^{3/2}}$$

Note: To find $\frac{d}{dy}\sqrt{64+16y^2}$, let $g = \sqrt{t} = \sqrt{64+16y^2}$ with $t = 64+16y^2$.

$$\frac{dg}{dy} = \frac{dg}{dt}\frac{dt}{dy} = \frac{d}{dt}\sqrt{t}\frac{d}{dy}(64+16y^2) = \frac{1}{2\sqrt{t}}(32y) = \frac{32y}{2\sqrt{64+16y^2}} = \frac{16y}{\sqrt{64+16y^2}}$$

Also, when this derivative was multiplied by y, it became $\frac{16y^2}{\sqrt{64+16y^2}}$.

---

[7] If you don't have this identity memorized, you can figure it out by drawing a right triangle. We want $\sin(\tan^{-1} u)$. The angle of the sine function equals $\tan^{-1} u$, so the tangent of that angle equals u. Draw a right triangle where the opposite is u and the adjacent is 1 (so the tangent is $\frac{u}{1} = u$). The hypotenuse will be $\sqrt{1+u^2}$. The sine is opposite over hypotenuse: $\frac{u}{\sqrt{1+u^2}}$. That's one way to find that $\sin(\tan^{-1} u) = \frac{u}{\sqrt{1+u^2}}$.

Techniques of Integration Calculus Practice Workbook

Example 11. Find the solution to the indefinite integral below (where $|x| > 1$).

$$\int \frac{dx}{\sqrt{x^2 - 1}}$$

Solution: According to the last bullet point on page 36, if we see an expression of the form $x^2 - a^2$, we should try[8] $x = a \sec \theta$ and $dx = a \sec \theta \tan \theta \, d\theta$. Compare $x^2 - 1$ with $x^2 - a^2$ to see that $a = 1$. Let $x = \sec \theta$ such that $dx = \sec \theta \tan \theta \, d\theta$. We will use the identity $\tan^2 \theta + 1 = \sec^2 \theta$. Subtract 1 from both sides to get $\tan^2 \theta = \sec^2 \theta - 1$. This allows us to replace $\sec \theta^2 - 1$ with $\tan^2 \theta$. Take the inverse secant of both sides of $x = \sec \theta$ to find that $\sec^{-1} x = \theta$. We'll use this expression to eliminate θ after integrating.

$$\int \frac{dx}{\sqrt{x^2 - 1}} = \int \frac{\sec \theta \tan \theta}{\sqrt{\sec \theta^2 - 1}} d\theta = \int \frac{\sec \theta \tan \theta}{\sqrt{\tan \theta^2}} d\theta = \int \frac{\sec \theta \tan \theta}{\tan \theta} d\theta = \int \sec \theta \, d\theta$$

$$= \ln|\sec \theta + \tan \theta| + c = \ln|x + \tan(\sec^{-1} x)| + c = \ln\left|x + \sqrt{x^2 - 1}\right| + c$$

Note: We used the trig identity[9] $\tan(\sec^{-1} x) = \sqrt{x^2 - 1}$.

Check the answer: Take a derivative of the answer with respect to x and verify that this matches the original integrand. Apply the chain rule (Example 1) repeatedly.

$$\frac{d}{dx}\left[\ln\left(x + \sqrt{x^2 - 1}\right) + c\right] = \frac{1}{x + \sqrt{x^2 - 1}} \frac{d}{dx}\left(x + \sqrt{x^2 - 1}\right)$$

$$= \frac{1}{x + \sqrt{x^2 - 1}}\left(1 + \frac{d}{dx}\sqrt{x^2 - 1}\right) = \frac{1}{x + \sqrt{x^2 - 1}}\left[1 + \frac{1}{2\sqrt{x^2 - 1}} \frac{d}{dx}(x^2 - 1)\right]$$

$$= \frac{1}{x + \sqrt{x^2 - 1}}\left(1 + \frac{2x}{2\sqrt{x^2 - 1}}\right) = \frac{1}{x + \sqrt{x^2 - 1}}\left(1 + \frac{x}{\sqrt{x^2 - 1}}\right)$$

$$= \frac{1}{x + \sqrt{x^2 - 1}}\left(\frac{\sqrt{x^2 - 1}}{\sqrt{x^2 - 1}} + \frac{x}{\sqrt{x^2 - 1}}\right) = \frac{1}{x + \sqrt{x^2 - 1}} \frac{\sqrt{x^2 - 1} + x}{\sqrt{x^2 - 1}} = \frac{1}{\sqrt{x^2 - 1}}$$

---

[8] Why? Since $\tan^2 \theta + 1 = \sec^2 \theta$, it follows that $\tan^2 \theta = \sec^2 \theta - 1$. So, for example, if we see the expression $x^2 - 1$, if we make the substitution $x = \sec \theta$, we get $x^2 - 1 = \sec^2 \theta - 1 = \tan^2 \theta$, which condenses the two terms $x^2 - 1$ down to the single term $\tan^2 \theta$.

[9] If you don't have this identity memorized, you can figure it out by drawing a right triangle. We want $\tan(\sec^{-1} x)$. The angle of the tangent function equals $\sec^{-1} x$, so the secant of that angle equals x. Draw a right triangle where the hypotenuse is x and the adjacent is 1 (so the secant, which is hypotenuse over adjacent, is $\frac{x}{1} = x$). Use the Pythagorean theorem to find the opposite side: $\sqrt{x^2 - 1}$. The tangent is opposite over adjacent: $\frac{\sqrt{x^2-1}}{1} = \sqrt{x^2 - 1}$. That's one way to find that $\tan(\sec^{-1} x) = \sqrt{x^2 - 1}$.

Chapter 3 Problems

Directions: Use a suitable substitution to perform each integral.

❶ $\displaystyle\int \tan(2\theta)\, d\theta$ Note: Like Example 1, don't use the antiderivative of tangent.

❷ $\displaystyle\int \frac{\cos x}{1 + \sin x}\, dx$

❸ $\displaystyle\int_{\varphi=0}^{\pi/6} \sin^2 \varphi \cos \varphi\, d\varphi$

❹ $\displaystyle\int_{\theta=0}^{\pi} \cos^2 \theta\, d\theta$

5 $\int \cos^3 x \, dx$

6 $\int \sin^2 y \cos^2 y \, dy$

7 $\int_{t=\pi/6}^{\pi} \sin^3 t \cos t \, dt$

8 $\int_{\theta=0}^{\pi/3} \sin\theta \sec^2\theta \, d\theta$

Chapter 3 – Trigonometric Substitutions

⑨ $\displaystyle\int \tan^2 \varphi \, d\varphi$

⑩ $\displaystyle\int \frac{\sec x \tan x}{\sqrt{\cos x}} \, dx$

⑪ $\displaystyle\int_{\theta=0}^{\pi/3} \sec^4 \theta \tan \theta \, d\theta$

⑫ $\displaystyle\int_{w=0}^{\pi/4} \tan^3 w \sec^2 w \, dw$

⑬ $\displaystyle\int \sqrt{16-x^2}\,dx \quad (|x|<4)$

⑭ $\displaystyle\int \frac{dt}{t^2+3}$

⑮ $\displaystyle\int_{y=2/\sqrt{3}}^{2} \frac{dy}{y^2-1}$

⑯ $\displaystyle\int_{x=0}^{5} \frac{x^2}{x^2+25}\,dx$

Chapter 3 – Trigonometric Substitutions

17 $\displaystyle\int \frac{\cos\theta}{\sqrt{1+\sin^2\theta}}\,d\theta$

18 $\displaystyle\int \frac{t}{(1-t^4)^{3/2}}\,dt \quad (|t|<1)$

19 $\displaystyle\int_{x=0}^{\pi/2} \frac{\sin(2x)}{1+\sin^2 x}\,dx$

20 $\displaystyle\int_{y=0}^{1} \frac{\tan^{-1} y}{1+y^2}\,dy$

4 Logarithmic and Power Substitutions

The integrals or substitutions in this chapter involve logarithms, exponentials, variable powers, hyperbolic functions, and their identities. So first we will quickly review these functions and some of their helpful identities.

A **logarithm** of base b has the form $\log_b x = y$ and corresponds to $b^y = x$. It basically asks, "Which exponent can you raise the base b to and obtain the argument x as a result?" For example, $\log_3 81 = y$ asks, "Which power of 3 equals 81?" The answer is $y = 4$ because $3^4 = 81$. The two most common logarithms are the base-10 logarithm $\log_{10} x$ and the **natural logarithm** $\ln x$ where the base equals **Euler's number** ($e \approx 2.71828...$) The natural logarithm equation $\ln x = y$ corresponds to the **exponential** equation $e^y = x$. An expression of the form e^y is an exponential. Logarithms and exponentials are related to one another.

Some common logarithm identities include:

$$\ln p + \ln q = \ln(pq) \quad , \quad \ln\left(\frac{p}{q}\right) = \ln p - \ln q \quad , \quad c \ln p = \ln p^c \quad , \quad -\ln p = \ln\frac{1}{p}$$

The **change of base** formula for logarithms is given below. This formula is handy when you need to find a logarithm that doesn't have a base of e or 10 using a calculator that only has $\ln$ and $\log_{10}$ buttons. For example, if you want to find log base 3 of 20, written as $\log_3 20$, which means which exponent of 3 will make 20, you could find this with the natural log button as $\frac{\ln 20}{\ln 3}$ or the base-ten logarithm button as $\frac{\log_{10} 20}{\log_{10} 3}$. Either way, a calculator would tell you that the answer is approximately 2.73.

$$\log_b x = \frac{\ln x}{\ln b} = \frac{\log_{10} x}{\log_{10} b}$$

Some common identities involving exponentials and other variable powers include:

$$\ln(e^x) = x \quad , \quad e^{\ln x} = x \quad , \quad e^{x \ln b} = b^x \quad , \quad \ln(e) = 1 \quad , \quad e^0 = 1 \quad , \quad \ln(1) = 0$$

$$b^x b^y = b^{x+y} \quad , \quad \frac{b^x}{b^y} = b^{x-y} \quad , \quad (b^x)^y = b^{xy}$$

$$b^{-x} = \frac{1}{b^x} \quad , \quad b^{1/x} = \sqrt[x]{b} \quad , \quad b^0 = 1 \ (b \neq 0) \quad , \quad b^{x/y} = (b^x)^{1/y} = \left(b^{1/y}\right)^x$$

Chapter 4 – Logarithmic and Power Substitutions

The **hyperbolic functions** include an important 'h' that distinguishes them from the ordinary trig functions. They are defined in terms of exponentials as follows.

$$\cosh x = \frac{e^x + e^{-x}}{2} \quad , \quad \sinh x = \frac{e^x - e^{-x}}{2} \quad , \quad \tanh x = \frac{\sinh x}{\cosh x} = \frac{e^x - e^{-x}}{e^x + e^x}$$

$$\operatorname{sech} x = \frac{1}{\cosh x} \quad , \quad \operatorname{csch} x = \frac{1}{\sinh x} \quad , \quad \coth x = \frac{1}{\tanh x}$$

Some common hyperbolic identities include:

$$\cosh x + \sinh x = e^x \quad , \quad \cosh x - \sinh x = e^{-x}$$

$$\cosh(x + y) = \cosh x \cosh y + \sinh x \sinh y$$

$$\sinh(x + y) = \sinh x \cosh y + \cosh x \sinh y$$

$$\tanh(x + y) = \frac{\tanh x + \tanh y}{1 + \tanh x \tanh y} \quad , \quad \tanh(2x) = \frac{2 \tanh x}{1 + \tanh^2 x} \quad \tanh\left(\frac{x}{2}\right) = \frac{\sinh x}{1 + \cosh x}$$

$$\cosh(2x) = \cosh^2 x + \sinh^2 x \quad , \quad \sinh(2x) = 2 \sinh x \cosh x$$

$$\cosh^2 x - \sinh^2 x = 1 \quad , \quad 1 - \tanh^2 x = \operatorname{sech}^2 x \quad , \quad \coth^2 x - 1 = \operatorname{csch}^2 x$$

$$(\cosh x + \sinh x)^n = e^{nx} = \cosh(nx) + \sinh(nx)$$

$$\tanh(\ln x) = \frac{x^2 - 1}{x^2 + 1} \quad , \quad e^{2x} = \frac{1 + \tanh x}{1 - \tanh x}$$

The **inverse hyperbolic functions** are related to natural logarithms as follows:

$$\cosh^{-1} x = \ln\left(x + \sqrt{x^2 - 1}\right) \quad x \geq 1, \quad \sinh^{-1} x = \ln\left(x + \sqrt{x^2 + 1}\right) \quad \text{all } x$$

$$\tanh^{-1} x = \frac{1}{2} \ln\left(\frac{1 + x}{1 - x}\right) \quad |x| < 1, \quad \coth^{-1} x = \frac{1}{2} \ln\left(\frac{x + 1}{x - 1}\right) \quad |x| > 1$$

$$\operatorname{sech}^{-1} x = \ln\left(\frac{1}{x} + \sqrt{\frac{1}{x^2} - 1}\right) \quad 0 < x \leq 1, \quad \operatorname{csch}^{-1} x = \ln\left(\frac{1}{x} + \sqrt{\frac{1}{x^2} + 1}\right) \quad x \neq 0$$

The hyperbolic trig functions and ordinary trig functions are related by the **imaginary number** i, defined[10] by $i^2 = -1$:

$$e^{iz} = \cos z + i \sin z \quad , \quad e^{inz} = \cos(nz) + i \sin(nz)$$

$$\cosh(ix) = \cos x \quad , \quad \sinh(ix) = i \sin x \quad , \quad \cos(ix) = \cosh x \quad , \quad \sin(ix) = i \sinh(x)$$

Recall the basic derivatives of the natural logarithm, exponential function, and the hyperbolic functions listed below.

---

[10] See Chapters 7-16 in Complex Numbers Essentials Math Workbook with Answers by Chris McMullen, Ph.D.

Techniques of Integration Calculus Practice Workbook

$$\frac{d}{dx}e^{ax} = ae^{ax} \quad , \quad \frac{d}{dx}b^x = b^x \ln b \quad , \quad \frac{d}{dx}\ln x = \frac{1}{x}$$

$$\frac{d}{dx}\sinh x = \cosh x \quad , \quad \frac{d}{dx}\cosh x = \sinh x$$

$$\frac{d}{dx}\tanh x = \text{sech}^2 x \quad , \quad \frac{d}{dx}\coth x = -\text{csch}^2 x$$

$$\frac{d}{dx}\text{sech}\, x = -\text{sech}\, x \tanh x \quad , \quad \frac{d}{dx}\text{csch}\, x = -\text{csch}\, x \coth x$$

The basic antiderivatives of these functions are given below. Since the exponential e^x is special in that it's the only function that equals its own derivative, it follows that the antiderivative of e^x also equals itself (apart from a constant of integration). In contrast, it's tricky to find the antiderivative of the natural logarithm; you can find the result below, but we'll wait until Chapter 7 to show how to derive it. The hyperbolic antiderivatives are similar to the ordinary trig antiderivatives (apart from a couple of sign differences), but hyperbolic secant is notable because its antiderivative can be expressed in terms of an inverse tangent (Example 5).

$$\int e^{ax} dx = \frac{e^{ax}}{a} + c \quad , \quad \int b^x dx = \frac{b^x}{\ln b} + c \ (b > 0, \neq 1), \quad \int \ln x\, dx = x\ln x - x + c$$

$$\int \cosh x\, dx = \sinh x + c \quad , \quad \int \sinh x\, dx = \cosh x + c$$

$$\int \tanh x\, dx = \ln(\cosh x) + c \quad , \quad \int \coth x\, dx = \ln|\sinh x| + c$$

$$\int \text{sech}\, x\, dx = \tan^{-1}|\sinh x| + c = 2\tan^{-1} e^{ax} \quad , \quad \int \text{csch}\, x\, dx = \ln\left|\tanh\frac{x}{2}\right| + c$$

Substitutions involving logarithms, exponentials, and hyperbolic functions work very much like the substitutions of Chapter 2-3. As before, it will help to be familiar with a variety of identities.

Example 1. Find the solution to the indefinite integral below.

$$\int \frac{e^x}{1 + e^x} dx$$

Solution: Since the numerator is the derivative of the denominator, let $u = 1 + e^x$ such that $du = e^x dx$.

Chapter 4 – Logarithmic and Power Substitutions

$$\int \frac{e^x}{1+e^x}\,dx = \int \frac{du}{u} = \ln|u| + c = \ln|1+e^x| + c = \ln(1+e^x) + c$$

Note: Since e^x is always positive, we don't need the absolute value symbols.

Check the answer: Take a derivative of the answer with respect to x and verify that this matches the original integrand. Apply the chain rule (see the previous chapters) with $f = \ln u + c = \ln(1+e^x) + c$ and $u = 1 + e^x$.

$$\frac{df}{dx} = \frac{df}{du}\frac{du}{dx} = \frac{d}{du}(\ln u + c)\frac{d}{dx}(1+e^x) = \frac{1}{u}e^x = \frac{e^x}{1+e^x}$$

Example 2. Perform the definite integral below.

$$\int_{t=1}^{e} \frac{\ln t}{t}\,dt$$

Solution: Since $\frac{1}{t}$ is the derivative of the natural logarithm, let $u = \ln t$ such that $du = \frac{dt}{t}$. Plug the given limits (1 and e) into the equation $u = \ln t$ to find the new limits of integration: $\ln(1) = 0$ and $\ln(e) = 1$.

$$\int_{t=1}^{e} \frac{\ln t}{t}\,dt = \int_{u=0}^{1} u\,du = \left[\frac{u^2}{2}\right]_{u=0}^{1} = \frac{1^2}{2} - \frac{0^2}{2} = \frac{1}{2} - 0 = \frac{1}{2} = 0.5$$

Check the antiderivative: Take a derivative with respect to t. Apply the chain rule with $f = \frac{u^2}{2} = \frac{(\ln t)^2}{2}$ and $u = \ln t$.

$$\frac{df}{dt} = \frac{df}{du}\frac{du}{dt} = \frac{d}{du}\frac{u^2}{2}\frac{d}{dt}\ln t = u\frac{1}{t} = \frac{\ln t}{t}$$

Example 3. Find the solution to the indefinite integral below.

$$\int \cosh y\,dy$$

Solution: Use the definition of hyperbolic cosine.

$$\int \cosh y\,dy = \int \frac{e^y + e^{-y}}{2}\,dy = \frac{e^y - e^{-y}}{2} + c = \sinh y + c$$

Check the answer: Take a derivative of the answer with respect to y and verify that this matches the original integrand: $\frac{d}{dy}(\sinh y + c) = \cosh y$.

Techniques of Integration Calculus Practice Workbook

Example 4. Perform the definite integral below.

$$\int_{w=0}^{1} \tanh w \, dw$$

Solution: Hyperbolic tangent equals hyperbolic sine over hyperbolic cosine.

$$\int_{w=0}^{1} \tanh w \, dw = \int_{w=0}^{1} \frac{\sinh w}{\cosh w} dw$$

Let $u = \cosh w$ such that $du = \sinh w \, dw$. (Unlike the ordinary trig functions, the derivative of hyperbolic cosine is positive hyperbolic sine.) Plug the given limits (0 and 1) into the equation $u = \cosh w$ to find the new limits of integration: $\cosh 0 = \frac{e^0 + e^{-0}}{2} = \frac{1+1}{2} = \frac{2}{2} = 1$ and $\cosh 1 = \frac{e^1 + e^{-1}}{2} \approx 1.543$.

$$\int_{w=0}^{1} \frac{\sinh w}{\cosh w} dw \approx \int_{u=1}^{1.543} \frac{du}{u} = [\ln|u|]_{u=1}^{1.543} = \ln 1.543 - \ln 1 = \ln 1.543 \approx 0.438$$

Check the antiderivative: Take a derivative with respect to w. Apply the chain rule with $f = \ln u = \ln \cosh w$ and $u = \cosh w$.

$$\frac{df}{dw} = \frac{df}{du}\frac{du}{dw} = \frac{d}{du}\ln u \frac{d}{dw}\cosh w = \frac{1}{u}\sinh w = \frac{\sinh w}{\cosh w} = \tanh w$$

Note: $\int \tanh w = \ln(\cosh w) + c$. No absolute values are needed because $\cosh w > 0$.

Example 5. Find the solution to the indefinite integral below.

$$\int \text{sech} \, x \, dx$$

Solution: Recall from Example 2 in Chapter 3 that the antiderivative of secant is very tricky. You could try to integrant hyperbolic secant the same way, but there turns out to be an easier way. We'll rewrite hyperbolic secant in terms of exponentials.[11]

$$\int \frac{2}{e^x + e^{-x}} dx$$

Wait a minute. Wasn't this supposed to make the integral easier? It may not look any easier. Well, it will be if we multiply the numerator and denominator each by e^x.

---

[11] You could use this method for $\int \sec \theta \, d\theta$ using complex exponentials. Since $e^{i\theta} = \cos \theta + i \sin \theta$, it follows that $\cos \theta = \frac{e^{i\theta} + e^{-i\theta}}{2}$. Ordinary secant is the reciprocal of this expression.

Chapter 4 – Logarithmic and Power Substitutions

$$\int \frac{2}{e^x + e^{-x}} dx = \int \frac{2}{e^x + e^{-x}} \frac{e^x}{e^x} dx = \int \frac{2e^x}{e^x e^x + e^{-x} e^x} dx = 2 \int \frac{e^x}{e^{2x} + 1} dx$$

Note that $e^x e^x = e^{x+x} = e^{2x}$ and $e^x e^{-x} = e^0 = 1$. Let $u = e^x$ such that $du = e^x dx$. Note that $u^2 = (e^x)^2 = e^{2x}$ according to the rule $(e^m)^n = e^{mn}$.

$$2 \int \frac{e^x}{e^{2x} + 1} dx = 2 \int \frac{du}{u^2 + 1}$$

Recall from Chapter 3 that if we see an expression of the form $u^2 + 1$, we should try the substitution $u = \tan \theta$ and $du = \sec^2 \theta \, d\theta$.

$$2 \int \frac{du}{u^2 + 1} = 2 \int \frac{\sec^2 \theta}{\tan^2 \theta + 1} d\theta = 2 \int \frac{\sec^2 \theta}{\sec^2 \theta} d\theta = 2 \int d\theta$$
$$= 2\theta + c = 2 \tan^{-1} u + c = 2 \tan^{-1}(e^x) + c$$

Check the answer: Take a derivative of the answer with respect to x and verify that this matches the original integrand. Apply the chain rule with $f = 2 \tan^{-1} u + c = 2 \tan^{-1}(e^x) + c$ and $u = e^x$. Recall from first-semester calculus that $\frac{d}{du} \tan^{-1} u = \frac{1}{1+u^2}$ provided that $-\frac{\pi}{2} < \tan^{-1} u < \frac{\pi}{2}$.

$$\frac{df}{dx} = \frac{df}{du} \frac{du}{dx} = \frac{d}{du}(2 \tan^{-1} u + c) \frac{d}{dx} e^x = \frac{2}{1+u^2} e^x = \frac{2e^x}{1+(e^x)^2}$$
$$= \frac{2e^x}{1 + e^{2x}} = \frac{2e^x}{1 + e^{2x}} \frac{e^{-x}}{e^{-x}} = \frac{2}{e^{-x} + e^x} = \text{sech } x$$

Example 6. Perform the definite integral below.

$$\int_{t=0}^{1} 2^t \, dt$$

Solution: Use the identity $e^{t \ln 2} = 2^t$. This identity works because $2 = e^{\ln 2}$. If we raise both sides to t, we get $2^t = (e^{\ln 2})^t = e^{t \ln 2}$ according to the rule $(e^m)^n = e^{mn}$.

$$\int_{t=0}^{1} 2^t \, dt = \int_{t=0}^{1} e^{t \ln 2} \, dt = \left[\frac{e^{t \ln 2}}{\ln 2}\right]_{t=0}^{1} = \left[\frac{2^t}{\ln 2}\right]_{t=0}^{1} = \frac{2^1}{\ln 2} - \frac{2^0}{\ln 2} = \frac{2}{\ln 2} - \frac{1}{\ln 2}$$
$$= \frac{2-1}{\ln 2} = \frac{1}{\ln 2} \approx 1.44$$

Check the antiderivative: $\frac{d}{dt} \frac{2^t}{\ln 2} = 2^t$.

Chapter 4 Problems

Directions: Use a suitable substitution to perform each integral.

❶ $\displaystyle\int e^x \sinh(e^x)\, dx$

❷ $\displaystyle\int \frac{\sqrt{1+\ln t}}{t}\, dt$

❸ $\displaystyle\int_{w=1}^{3} \frac{5^w}{\sqrt{5^w - 4}}\, dw$

❹ $\displaystyle\int_{y=1/e}^{1} \ln\left(\frac{1}{y}\right) dy$

Chapter 4 – Logarithmic and Power Substitutions

❺ $\int \coth z \, dz$ Note: Like Example 4, don't use the antiderivative of coth t.

❻ $\int \tanh^{-1} y \, dy$

❼ $\int_{t=0}^{1} \sinh t \, dt$ Note: Like Example 3, don't use the antiderivative of sinh t.

❽ $\int_{x=0}^{1} \cosh^2 x \, dx$

⑨ $\int \dfrac{2\tanh z}{1+\tanh^2 z}\,dz$

⑩ $\int (\cosh y + \sinh y)^3\,dy$

⑪ $\displaystyle\int_{x=0}^{1} \tanh x\,\operatorname{sech}^2 x\,dx$

⑫ $\displaystyle\int_{w=0}^{1} \dfrac{1+\tanh w}{1-\tanh w}\,dw$

5 Completing the Square

First, we will discuss the algebraic technique known as completing the square. (We'll also provide a handy formula for it.) Then we'll discuss how it can help with integrals.

Consider the quadratic $9x^2 + 24x + 5$. It 'almost' looks like a perfect square. The leading term, $9x^2$, is the square of $3x$. That is, $(3x)^2 = 3^2x^2 = 9x^2$. The second term, $24x$, would be the cross term if we squared $3x + 4$, since the cross term would be $3x(4) + 4(3x) = 12x + 12x = 24x$. The only problem is the constant, 5. If we square $3x + 4$, we get a different constant: $(3x + 4)(3x + 4) = 9x^2 + 24x + 16$. Compare the quadratic $9x^2 + 24x + 5$ with the perfect square $(3x + 4)^2 = 9x^2 + 24x + 16$. The only difference is that the constant is 5 instead of 16. In a sense, the quadratic $9x^2 + 24x + 5$ is 'almost' a perfect square; it's 'incomplete.' If the constant had been 16 instead of 5, it would be 'complete'; it would be a 'perfect' square.

The main idea behind **completing the square** is to shift the constant term of a quadratic expression in order to make it a perfect square. We'll illustrate this with the previous example. If we add $11 - 11 = 0$ to the quadratic $9x^2 + 24x + 5$, it is transformed into $9x^2 + 24x + 5 + 11 - 11$, which we may rewrite as $9x^2 + 24x + 16 - 11$. The first three terms are now a perfect square, so we can write it as $(3x + 4)^2 - 11$. You can check that this works by multiplying it out:
$$(3x + 4)^2 - 11 = 9x^2 + 24x + 16 - 11 = 9x^2 + 24x + 5$$
When we rewrite $9x^2 + 24x + 5$ in the form $(3x + 4)^2 - 11$, we say that we have **'completed the square.'**

It turns out that **any** quadratic (that isn't already a perfect square) can have its square completed in this manner. Let's do one more example. Consider $16x^2 - 40x + 32$. Its square can be completed two different ways. One way is to think through the steps logically as illustrated below. The second way is apply a formula. If you have trouble applying this strategy by thinking through the steps, all you need to do is memorize the formula (that we'll provide shortly), and you'll be able to skip these steps. The main idea is to think of what you could square to form $16x^2 - 40x +$ a constant.

- What could you square to make the leading term, $16x^2$? The answer is $4x$ since $(4x)^2 = 4^2 x^2 = 16x^2$.
- What could you add (or subtract) to $4x$ (which we found in the previous bullet point) to make the cross term, which is $-40x$? We want to square $(4x + k)$ so that it makes $4x^2 - 40x + $ a constant. If you think about squaring $(4x + k)$, the cross term will $8kx$ (since we would get $4xk + k4x$); it's 2 times $4x$ times k. Since the cross term is $-40x$, and this equals $8kx$, we need k to be -5. That is, if we square $(4x - 5)$, we'll get $16x^2 - 40x + $ a constant.
- Square $4x - 5$, which we found in the previous step, to figure out what that constant should be: $(4x - 5)^2 = 16x^2 - 40x + 25$.
- What do we need to do to the given expression, $16x^2 - 40x + 32$, to make the perfect square in the previous step? We need to subtract and add seven. That is, $16x^2 - 40x + 32 - 7 + 7 = 16x^2 - 40x + 25 + 7$.
- Rewrite the answer to the previous step as a perfect square plus (or minus) a constant: $16x^2 - 40x + 25 + 7 = (4x - 5)^2 + 7$.
- Check the answer: $(4x - 5)^2 + 7 = 16x^2 - 40x + 25 + 7 = 16x^2 - 40x + 32$.

This may seem like a lot of work, but we showed a lot more work than necessary (plus all the reasoning). It can be done concisely. But you don't have to go through those steps if you don't want to. You can use the following formula instead.

Given a quadratic expression in the standard form $ax^2 + bx + c$, if you identify the coefficients a, b, and c, you may use these coefficients to complete the square using the **formula** below:

$$ax^2 + bx + c = \left(x\sqrt{a} + \frac{b}{2\sqrt{a}}\right)^2 + \left(c - \frac{b^2}{4a}\right)$$

Let's apply this formula to the previous example to verify that it works for that case. Compare $16x^2 - 40x + 32$ with $ax^2 + bx + c$ to identify $a = 16, b = -40$, and $c = 32$. Plug these into the formula.

$$\left(x\sqrt{16} + \frac{-40}{2\sqrt{16}}\right)^2 + \left[32 - \frac{(-40)^2}{4(16)}\right] = \left[4x - \frac{40}{2(4)}\right]^2 + \left(32 - \frac{1600}{64}\right)$$
$$= (4x - 5)^2 + (32 - 25) = (4x - 5)^2 + 7$$

Chapter 5 – Completing the Square

There is an **important exception** to the formula: If the coefficient of the x^2 term is negative, like $3 + 6x - 9x^2$, **factor out an overall minus sign** before using the formula for completing the square. In this case, we get $(-1)(-3 - 6x + 9x^2)$. In this example, we also need to reorder the terms, since they need to go from highest power to lowest power before identifying the coefficients: $(-1)(9x^2 - 6x - 3)$. After factoring out the (-1), the coefficients are $a = 9$, $b = -6$, and $c = -3$. Use the formula.

$$(-1)\left\{\left(x\sqrt{9} + \frac{-6}{2\sqrt{9}}\right)^2 + \left[-3 - \frac{(-6)^2}{4(9)}\right]\right\} = (-1)\left\{\left[3x - \frac{6}{2(3)}\right]^2 + \left(-3 - \frac{36}{36}\right)\right\}$$

$$= (-1)[(3x-1)^2 + (-3-1)] = (-1)[(3x-1)^2 - 4] = 4 - (3x-1)^2$$

As always, you can check the answer by multiplying it out:

$$4 - (3x-1)^2 = 4 - (9x^2 - 6x + 1) = 4 - 9x^2 + 6x - 1 = -9x^2 + 6x + 3$$

How does completing the square help with integration? Consider $\int \frac{dx}{x^2+2x+5}$. If the denominator were $x^2 + 2x + 1$, this would be easy, since $x^2 + 2x + 1 = (x+1)^2$. The integral would then be $\int \frac{dx}{(x+1)^2}$, and we would let $u = x + 1$ like we did in Chapter 2. But the denominator is $x^2 + 2x + 5$ instead of $x^2 + 2x + 1$. However, if we complete the square, we'll be able to do the integral. If we complete the square, we get

$$x^2 + 2x + 5 = x^2 + 2x + 1 + 4 = (x+1)^2 + 4$$

This allows us to rewrite the integral $\int \frac{dx}{x^2+2x+5}$ as $\int \frac{dx}{(x+1)^2+4}$. How does this help? If we let $u = x + 1$ and $du = dx$, it becomes $\int \frac{du}{u^2+4}$. Now we can solve the integral using a suitable trigonometric substitution from Chapter 3. See Example 4.

The idea is that if an integral contains a quadratic expression like $ax^2 + bx + c$, after completing the square we may be able to perform the integral with a trigonometric substitution. This strategy is illustrated in Examples 4-5.

Techniques of Integration Calculus Practice Workbook

Example 1. Complete the square for the quadratic below.
$$36x^2 - 132x + 45$$
Solution: One way is to use the formula with $a = 36$, $b = -132$, and $c = 45$.
$$\left(x\sqrt{a} + \frac{b}{2\sqrt{a}}\right)^2 + \left(c - \frac{b^2}{4a}\right) = \left(x\sqrt{36} + \frac{-132}{2\sqrt{36}}\right)^2 + \left[45 - \frac{(-132)^2}{4(36)}\right]$$
$$= \left[6x - \frac{132}{2(6)}\right]^2 + \left(45 - \frac{17{,}424}{144}\right) = (6x - 11)^2 + (45 - 121) = (6x - 11)^2 - 76$$
Check the answer: $(6x - 11)^2 - 76 = 36x^2 + 2(6x)(-11) + 121 - 76 = 36x^2 - 132x + 45$

Example 2. Complete the square for the quadratic below.
$$5x^2 + x\sqrt{5} + 10$$
Solution: One way is to use the formula with $a = 5$, $b = \sqrt{5}$, and $c = 10$.
$$\left(x\sqrt{a} + \frac{b}{2\sqrt{a}}\right)^2 + \left(c - \frac{b^2}{4a}\right) = \left[x\sqrt{5} + \frac{(\sqrt{5})^2}{2\sqrt{5}}\right]^2 + \left[10 - \frac{(\sqrt{5})^2}{4(5)}\right]$$
$$= \left(x\sqrt{5} + \frac{\sqrt{5}}{2\sqrt{5}}\right)^2 + \left(10 - \frac{5}{20}\right) = \left(x\sqrt{5} + \frac{1}{2}\right)^2 + \left(10 - \frac{1}{4}\right) = \left(x\sqrt{5} + \frac{1}{2}\right)^2 + \frac{39}{4}$$
Check the answer: $\left(x\sqrt{5} + \frac{1}{2}\right)^2 + \frac{39}{4} = 5x^2 + 2(x\sqrt{5})\left(\frac{1}{2}\right) + \left(\frac{1}{2}\right)^2 + \frac{39}{4}$
$$= 5x^2 + x\sqrt{5} + \frac{1}{4} + \frac{39}{4} = 5x^2 + x\sqrt{5} + \frac{40}{4} = 5x^2 + x\sqrt{5} + 10$$

Example 3. Complete the square for the quadratic below.
$$15 - 24x - 16x^2$$
Solution: Since the x^2 term is negative, factor out an overall minus sign. Also, before identifying the coefficients, reorder the terms from highest power to lowest power: $(-1)(-15 + 24x + 16x^2) = (-1)(16x^2 + 24x - 15)$. Identify the coefficients: $a = 16$, $b = 24$, and $c = -15$. Use the formula (if you need it).
$$(-1)\left[\left(x\sqrt{a} + \frac{b}{2\sqrt{a}}\right)^2 + \left(c - \frac{b^2}{4a}\right)\right] = (-1)\left\{\left(x\sqrt{16} + \frac{24}{2\sqrt{16}}\right)^2 + \left[-15 - \frac{24^2}{4(16)}\right]\right\}$$
$$= (-1)\left\{\left[4x + \frac{24}{2(4)}\right]^2 + \left(-15 - \frac{576}{64}\right)\right\} = (-1)[(4x + 3)^2 + (-15 - 9)]$$
$$= (-1)[(4x + 3)^2 - 24] = 24 - (4x + 3)^2$$

Chapter 5 – Completing the Square

Check the answer: $24 - (4x+3)^2 = 24 - [16x^2 + 2(4x)(3) + 9]$
$24 - (16x^2 + 24x + 9) = 24 - 16x^2 - 24x - 9 = -16x^2 - 24x + 15$

Example 4. Find the solution to the indefinite integral below.
$$\int \frac{dx}{x^2 + 2x + 5}$$
Solution: Complete the square for $x^2 + 2x + 5$. Identify the coefficients: $a = 1$, $b = 2$, and $c = 5$. Use the formula (if you need it).
$$\left(x\sqrt{a} + \frac{b}{2\sqrt{a}}\right)^2 + \left(c - \frac{b^2}{4a}\right) = \left(x\sqrt{1} + \frac{2}{2\sqrt{1}}\right)^2 + \left[5 - \frac{2^2}{4(1)}\right]$$
$$= (x+1)^2 + \left(5 - \frac{4}{4}\right) = (x+1)^2 + (5-1) = (x+1)^2 + 4$$
Before we move on, let's check the answer for completing the square:
$$(x+1)^2 + 4 = x^2 + 2x + 1 + 4 = x^2 + 2x + 5$$
Rewrite the integral with the completed square.
$$\int \frac{dx}{x^2 + 2x + 5} = \int \frac{dx}{(x+1)^2 + 4}$$
Let $u = x + 1$ such that $du = dx$.
$$\int \frac{dx}{(x+1)^2 + 4} = \int \frac{du}{u^2 + 4}$$
You should recognize this as a trig substitution from Chapter 3. Let $u = 2\tan\theta$ such that $du = 2\sec^2\theta\, d\theta$.
$$\int \frac{du}{u^2 + 4} = \int \frac{2\sec^2\theta}{(2\tan\theta)^2 + 4} d\theta = \int \frac{2\sec^2\theta}{4\tan^2\theta + 4} d\theta = \int \frac{2\sec^2\theta}{4(\tan^2\theta + 1)} d\theta$$
$$= \frac{1}{2}\int \frac{\sec^2\theta}{\sec^2\theta} d\theta = \frac{1}{2}\int d\theta = \frac{\theta}{2} + c = \frac{1}{2}\tan^{-1}\left(\frac{u}{2}\right) + c = \frac{1}{2}\tan^{-1}\left(\frac{x+1}{2}\right) + c$$
Check the answer: Take a derivative of the answer with respect to x and verify that this matches the original integrand. Apply the chain rule (see the previous chapters).
$$\frac{d}{dx}\left[\frac{1}{2}\tan^{-1}\left(\frac{x+1}{2}\right) + c\right] = \frac{1}{2}\frac{1}{1+\left(\frac{x+1}{2}\right)^2}\frac{d}{dx}\left(\frac{x+1}{2}\right) = \frac{1}{2}\frac{1}{1+\left(\frac{x^2+2x+1}{4}\right)}\left(\frac{1}{2}\right)$$
$$= \frac{1}{4\left[1+\left(\frac{x^2+2x+1}{4}\right)\right]} = \frac{1}{4 + x^2 + 2x + 1} = \frac{1}{x^2 + 2x + 5}$$

Example 5. Perform the definite integral below.
$$\int_{x=-1.5}^{1} \frac{dx}{(91 - 12x - 4x^2)^{3/2}}$$

Solution: Complete the square for $91 - 12x - 4x^2$. Since the x^2 term is negative, factor out a minus sign (like we did in Example 3). Also, reorder the terms from highest power to lowest power: $(-1)(-91 + 12x + 4x^2) = (-1)(4x^2 + 12x - 91)$. Identify the coefficients: $a = 4$, $b = 12$, and $c = -91$. Use the formula.

$$(-1)\left[\left(x\sqrt{a} + \frac{b}{2\sqrt{a}}\right)^2 + \left(c - \frac{b^2}{4a}\right)\right] = (-1)\left\{\left(x\sqrt{4} + \frac{12}{2\sqrt{4}}\right)^2 + \left[-91 - \frac{12^2}{4(4)}\right]\right\}$$

$$= (-1)\left[\left[2x + \frac{12}{2(2)}\right]^2 + \left(-91 - \frac{144}{16}\right)\right] = (-1)[(2x+3)^2 + (-91 - 9)]$$

$$= (-1)[(2x+3)^2 - 100] = 100 - (2x+3)^2$$

Before we move on, let's check the answer for completing the square:
$$100 - (2x+3)^2 = 100 - (4x^2 + 12x + 9)$$
$$= 100 - 4x^2 - 12x - 9 = -4x^2 - 12x + 91$$

Rewrite the integral with the completed square.

$$\int_{x=-1.5}^{1} \frac{dx}{(91 - 12x - 4x^2)^{3/2}} = \int_{x=-1.5}^{1} \frac{dx}{[100 - (2x+3)^2]^{3/2}}$$

Let $u = 2x + 3$ such that $du = 2dx$, which equates to $\frac{du}{2} = dx$. When $x = -1.5$, $u = 2(-1.5) + 3 = -3 + 3 = 0$. When $x = 1$, $u = 2(1) + 3 = 2 + 3 = 5$.

$$\int_{u=0}^{5} \frac{du}{2(100 - u^2)^{3/2}} = \frac{1}{2}\int_{u=0}^{5} \frac{du}{(100 - u^2)^{3/2}} = \frac{1}{2}\int_{u=0}^{5} \frac{du}{(10^2 - u^2)^{3/2}}$$

You should recognize this as a trig substitution from Chapter 3. Let $u = 10\sin\theta$ such that $du = 10\cos\theta\, d\theta$. When $u = 0$, $\theta = \sin^{-1}\left(\frac{0}{10}\right) = 0$. When $u = 5$, $\theta = \sin^{-1}\left(\frac{5}{10}\right) = \sin^{-1}\left(\frac{1}{2}\right) = \frac{\pi}{6}$.

$$\frac{1}{2}\int_{u=0}^{5} \frac{du}{(10^2 - u^2)^{3/2}} = \frac{1}{2}\int_{\theta=0}^{\pi/6} \frac{10\cos\theta}{[10^2 - (10\sin\theta)^2]^{3/2}} d\theta$$

Chapter 5 – Completing the Square

$$= 5 \int_{\theta=0}^{\pi/6} \frac{\cos\theta}{(100-100\sin^2\theta)^{3/2}} d\theta = 5 \int_{\theta=0}^{\pi/6} \frac{\cos\theta}{[100(1-\sin^2\theta)]^{3/2}} d\theta$$

$$= 5 \int_{\theta=0}^{\pi/6} \frac{\cos\theta}{[100\cos^2\theta]^{3/2}} d\theta = 5 \int_{\theta=0}^{\pi/6} \frac{\cos\theta}{100^{3/2}\cos^3\theta} d\theta = \frac{5}{100^{3/2}} \int_{\theta=0}^{\pi/6} \frac{1}{\cos^2\theta} d\theta$$

$$= \frac{5}{1000} \int_{\theta=0}^{\pi/6} \sec^2\theta\, d\theta = \frac{1}{200} [\tan\theta]_{\theta=0}^{\pi/6} = \frac{1}{200}\left(\tan\frac{\pi}{6} - \tan 0\right)$$

$$= \frac{1}{200}\left(\frac{\sqrt{3}}{3} - 0\right) = \frac{\sqrt{3}}{600} \approx 0.00289$$

Notes: The integral $\int \sec^2\theta\, d\theta = \tan\theta + c$ is easy because $\frac{d}{d\theta}\tan\theta = \sec^2\theta$. (See Exercise 13 in Chapter 1.)

Check the antiderivative: The antiderivative is $\frac{1}{200}\tan\left[\sin^{-1}\left(\frac{u}{10}\right)\right] = \frac{1}{200}\frac{u/10}{\sqrt{1-\left(\frac{u}{10}\right)^2}} =$

$\frac{1}{200}\frac{u}{10\sqrt{1-\frac{u^2}{100}}} = \frac{1}{200}\frac{u}{\sqrt{100}\sqrt{1-\frac{u^2}{100}}} = \frac{1}{200}\frac{u}{\sqrt{100-u^2}}$ where $u = 2x + 3$. We used the trig identity[12]

$\tan(\sin^{-1} u) \frac{u}{\sqrt{1-u^2}}$. Take a derivative with respect to x and verify that this matches the original integrand. Use the chain rule. Also use the **quotient rule**.

$$\frac{df}{dx} = \frac{df}{du}\frac{du}{dx} = \frac{d}{du}\frac{1}{200}\frac{u}{\sqrt{100-u^2}}\frac{d}{dx}(2x+3) = \frac{\sqrt{100-u^2}\frac{d}{du}u - u\frac{d}{du}\sqrt{100-u^2}}{200(100-u^2)} \quad (2)$$

$$= \frac{\sqrt{100-u^2} - u\frac{1}{2\sqrt{100-u^2}}\frac{d}{du}(100-u^2)}{100(100-u^2)} = \frac{\sqrt{100-u^2} - u\frac{-2u}{2\sqrt{100-u^2}}}{100(100-u^2)}$$

$$= \frac{\sqrt{100-u^2} + \frac{u^2}{\sqrt{100-u^2}}}{100(100-u^2)} = \frac{\sqrt{100-u^2}\frac{\sqrt{100-u^2}}{\sqrt{100-u^2}} + \frac{u^2}{\sqrt{100-u^2}}}{100(100-u^2)}$$

---

[12] If you don't have this identity memorized, you can figure it out by drawing a right triangle. We want $\tan(\sin^{-1} u)$. The angle of the tangent function equals $\sin^{-1} u$, so the sine of that angle equals u. Draw a right triangle where the opposite is u and the hypotenuse is 1 (so the sine, which is opposite over hypotenuse, is $\frac{u}{1} = u$). Use the Pythagorean theorem to find the adjacent side: $\sqrt{1-u^2}$. The tangent is opposite over adjacent: $\frac{u}{\sqrt{1-u^2}}$. That's one way to find that $\tan(\sin^{-1} u) = \frac{u}{\sqrt{1-u^2}}$.

$$= \frac{\frac{100-u^2}{\sqrt{100-u^2}} + \frac{u^2}{\sqrt{100-u^2}}}{100(100-u^2)} = \frac{\frac{100-u^2+u^2}{\sqrt{100-u^2}}}{100(100-u^2)} = \frac{\frac{100}{\sqrt{100-u^2}}}{100(100-u^2)}$$

$$= \frac{100}{\sqrt{100-u^2}} \div [100(100-u^2)] = \frac{100}{\sqrt{100-u^2}} \times \frac{1}{100(100-u^2)}$$

$$= \frac{1}{(100-u^2)\sqrt{100-u^2}} = \frac{1}{(100-u^2)^{3/2}}$$

$$= \frac{1}{[100-(2x+3)^2]^{3/2}} = \frac{1}{[100-(4x^2+12x+9)]^{3/2}} = \frac{1}{(91-12x-4x^2)^{3/2}}$$

Chapter 5 Problems

Directions: Complete each square.

❶ $x^2 + 12x + 21$

❷ $4y^2 + 20y + 42$

❸ $25w^2 - 30w - 6$

❹ $75 - 96t - 36t^2$

❺ $81u^2 + 54u - 1$

❻ $16x^2 - 32x + 11$

❼ $50 - 42z^2 + 9z^4$

❽ $144y^2 - 48y$

❾ $121 + 140t - 49t^2$

❿ $12w^2 - 8w\sqrt{3} - 2$

⓫ $9x + 24\sqrt{x} + 2$

⓬ $4\cos^2\theta + 4\cos\theta + 5$

Directions: Complete a square to perform each integral.

⑬ $\displaystyle\int \frac{dx}{\sqrt{x^2 - 10x + 29}}$

⑭ $\displaystyle\int \frac{dy}{25y^2 - 10y - 8}$

⑮ $\displaystyle\int \frac{dp}{\sqrt{8+6p-9p^2}}$

⑯ $\displaystyle\int \frac{dw}{\sqrt{16w^2+24w}}$

⑰ $\displaystyle\int \frac{x}{4x^4 + 12x^2 + 25}\,dx$

⑱ $\displaystyle\int \frac{t}{t^2 + 6t + 5}\,dt$

19 $$\int_{y=5/3}^{10/3} \frac{dy}{\sqrt{9y^2 - 30y + 50}}$$

20 $$\int_{\theta=0}^{\pi/2} \frac{\cos\theta}{\cos^2\theta + 2\sin\theta + 2}\, d\theta$$

6 Partial Fractions

If the integrand is a fraction of polynomials, like the form shown below (including the case where q happens to be a constant), you can use a method called partial fractions to perform the integral.

$$\int \frac{q(x)}{p(x)} dx$$

If the degree of $p(x)$ is greater than the degree of $q(x)$, meaning that $p(x)$ has a higher power than $q(x)$, then the method of partial fractions helps us rewrite the ratio $\frac{q(x)}{p(x)}$ in terms of simpler fractions. If $p(x)$ doesn't have a higher degree than $q(x)$, we first divide $q(x)$ by $p(x)$ using polynomial division, and then apply the method of partial fractions to the remainder. See Example 3. (If you don't know how to do long division with polynomials, or if you've forgotten it, or if you just don't like it, don't worry. We will show you another way to divide polynomials.)

Consider the integral below. The denominator $(1 - x^2)$ has a higher degree than the numerator (which is simply the constant 1, since $dx = 1dx$). You should already know how to do this integral using trig substitution (Chapter 3), and now we will learn how to do it using partial fractions. Sometimes, the method of partial fractions is simpler.

$$\int \frac{dx}{1 - x^2}$$

First, note that this denominator factors as $1 - x^2 = (1 + x)(1 - x)$. We will use this factoring to rewrite the above fraction in terms of two simpler fractions, where these factors are the denominators of the new fractions. We don't yet know what the numerators of those fractions will be, so we'll call them a and b for now.

$$\frac{1}{1 - x^2} = \frac{a}{1 + x} + \frac{b}{1 - x}$$

To determine the constants a and b, make a common denominator. (Actually, all you really need to do is think about making a common denominator; we only need to work out what the numerators would be after adding the fractions.)

$$\frac{1}{1 - x^2} = \frac{a(1 - x)}{(1 + x)(1 - x)} + \frac{b(1 + x)}{(1 - x)(1 + x)} = \frac{a - ax + b + bx}{(1 + x)(1 - x)}$$

Compare the original fraction with the new fraction on the right where we made the common denominator. These two fractions must be equal. The denominators are equal by design. The numerators will be equal if the equation below is true.
$$1 = a - ax + b + bx$$
Factor out the x.
$$1 = a + b + x(b - a)$$
Now think about this. How could both sides of the equation be true for any possible value of x? If the coefficient of x were zero, that would help. The coefficient of x would be zero if $b - a = 0$, meaning that $a = b$. If we let $a = b$, the above equation becomes:
$$1 = b + b + 0 = 2b$$
This is solved by $b = \frac{1}{2}$. Since $a = b$, then $a = \frac{1}{2}$ also. The constants don't always turn out to be equal; it just worked out that way in this simple case. The examples will help illustrate this technique in general. Plug $a = b = \frac{1}{2}$ into the partial fractions.
$$\frac{1}{1-x^2} = \frac{1/2}{1+x} + \frac{1/2}{1-x}$$
Once we have the **partial fraction decomposition** (that's what the equation above is), we can use it to make the original integral simpler. (Assume that $|x| < 1$.)
$$\int \frac{dx}{1-x^2} = \frac{1}{2}\int \frac{dx}{1+x} + \frac{1}{2}\int \frac{dx}{1-x} = \frac{1}{2}\int \frac{du}{u} + \frac{1}{2}\int \frac{-dt}{t} = \frac{1}{2}\int \frac{du}{u} - \frac{1}{2}\int \frac{dt}{t}$$
We made the substitutions $u = 1 + x$, $du = dx$, $t = 1 - x$, and $dt = -dx$.
$$= \frac{\ln|u|}{2} - \frac{\ln|t|}{2} = \frac{1}{2}\ln\left|\frac{u}{t}\right| = \frac{1}{2}\ln\left|\frac{1+x}{1-x}\right| + c$$
In the last step, we used $\ln y - \ln z = \ln\left(\frac{y}{z}\right)$. We could also use $\frac{1}{2}\ln w = \ln w^{1/2} = \ln \sqrt{w}$ to express the answer as $\ln \sqrt{\frac{1+x}{1-x}} + c$. Following is a quick summary:

- We wrote the integrand $\frac{1}{1-x^2}$ as $\frac{a}{1+x} + \frac{b}{1-x}$.
- We used the process of finding a common denominator to solve for the constants a and b.
- We set the integral $\int \frac{dx}{1-x^2}$ equal to $\int \frac{a}{1+x}dx + \int \frac{b}{1-x}dx$.
- We did these two integrals instead of the original integral.

Chapter 6 – Partial Fractions

Example 1. Find the solution to the indefinite integral below.
$$\int \frac{dx}{3x^2 + 14x - 5}$$
Solution: Factor[13] the quadratic: $3x^2 + 14x - 5 = (3x - 1)(x + 5)$.
$$\int \frac{dx}{3x^2 + 14x - 5} = \int \frac{a}{3x - 1} dx + \int \frac{b}{x + 5} dx$$
Imagine[14] making a common denominator to add the fractions $\frac{a}{3x-1} + \frac{b}{x+5}$. To do this, you would multiply[15] a by $(x + 5)$ and b by $(3x - 1)$ to get $a(x + 5) + b(3x - 1)$. Since the original integrand has 1 in the numerator (since $dx = 1dx$), the new numerators would need to add up to 1 (after making the common denominator).
$$1 = a(x + 5) + b(3x - 1)$$
Distribute the constants a and b.
$$1 = ax + 5a + 3bx - b$$
Factor out the x.
$$1 = (a + 3b)x + 5a - b$$
How could this equation be true for any possible value of x?

- If $a + 3b = 0$, that's how the equation could be true regardless of what x is.
- In that case, we would also need $1 = 5a - b$.

We now have two equations and two unknown constants (a and b). Solve this system of equations to determine the values of a and b.

---

[13] Factoring is an art, and some students are better at it than others. Look at the possible factors of the 3 in $3x^2$ and the constant (-5). One factor of -5 will be positive and one will be negative. If it factors with integers, the choices are $(x - 1)(3x + 5)$, $(x + 1)(3x - 5)$, $(3x - 1)(x + 5)$, or $(3x + 1)(x - 5)$. Which of these multiply out to $3x^2 + 14x - 5$? The answer is $(3x - 1)(x + 5)$. Since the leading coefficient and constant coefficient are prime numbers in this example, there are only four combinations. When the coefficients aren't prime numbers, there can be many more combinations. If you struggle with factoring, for quadratic equations you can use the quadratic formula to figure out how it factors. In this case, the quadratic formula would tell you $x = \frac{1}{3}$ or $x = -5$, corresponding to $\left(x - \frac{1}{3}\right)(x + 5)$, and then if you multiply this by 3, you get $(3x - 1)(x + 5)$.

[14] If you don't want to imagine, get a pencil and paper and just do it: $\frac{a(x+5)}{(3x-1)(x+5)} + \frac{b(3x-1)}{(x+5)(3x-1)}$. Now it's easy to 'see' that the numerators add up to $a(x + 5) + b(3x - 1)$. We're only working out what the numerators will add up to because that's all we need to solve for a and b. Feel free to do more if you want.

[15] If you're familiar with the term 'cross multiply,' this may be a helpful way to think of it. (At least, when there are only two fractions to add together.)

$$a + 3b = 0$$
$$1 = 5a - b$$

Isolate a in the first equation: $a = -3b$. Substitute this into the second equation.
$$1 = 5(-3b) - b = -15b - b = -16b$$
$$-\frac{1}{16} = b$$

Plug the value for b into the previous equation $a = -3b$.
$$a = -3b = -3\left(-\frac{1}{16}\right) = \frac{3}{16}$$

Plug $a = \frac{3}{16}$ and $b = -\frac{1}{16}$ into the partial fraction decomposition of the integral.
$$\int \frac{dx}{3x^2 + 14x - 5} = \frac{3}{16}\int \frac{dx}{3x - 1} - \frac{1}{16}\int \frac{dx}{x + 5}$$

Let $u = 3x - 1$, $du = 3dx$, $\frac{du}{3} = dx$, $t = x + 5$, and $dt = dx$ in these integrals.
$$\frac{3}{16}\int \frac{dx}{3x-1} - \frac{1}{16}\int \frac{dx}{x+5} = \frac{3}{16}\int \frac{du}{3u} - \frac{1}{16}\int \frac{dt}{t} = \frac{1}{16}\int \frac{du}{u} - \frac{1}{16}\int \frac{dt}{t}$$
$$= \frac{\ln|u|}{16} - \frac{\ln|t|}{16} = \frac{\ln|3x-1|}{16} - \frac{\ln|x+5|}{16} + c = \frac{1}{16}\ln\left|\frac{3x-1}{x+5}\right| + c$$

Note: We used $\ln y - \ln z = \ln \frac{y}{z}$ in the last step.

Alternate answer: $\frac{1}{16}\ln\left|\frac{3x-1}{3x+15}\right| + c$. How? Obviously, $x + 5$ isn't the same as $3x + 15$. The 'trick' is that $\ln\left(\frac{z}{3}\right) = \ln z - \ln 3 = \ln z -$ a constant. Since $3x + 15 = 3(x + 5)$, we may write $\frac{1}{16}\ln\left|\frac{3x-1}{3x+15}\right| = \frac{1}{16}\ln\left|\frac{3x-1}{3(x+5)}\right|$. Now let $z = \frac{3x-1}{x+5}$. This allows us to use $\ln\left(\frac{z}{3}\right) = \ln z - \ln 3$ to write $\frac{1}{16}\ln\left|\frac{3x-1}{3(x+5)}\right| = \frac{1}{16}\ln\left|\frac{3x-1}{x+5}\right| +$ a constant.

Check the answer: Take a derivative of the answer with respect to x and verify that this matches the original integrand. Apply the chain rule (see the previous chapters).
$$\frac{d}{dx}\left[\frac{1}{16}\ln(3x-1) - \frac{1}{16}\ln(x+5) + c\right] = \frac{1}{16(3x-1)}\frac{d}{dx}(3x-1) - \frac{1}{16(x+5)}\frac{d}{dx}(x+5)$$
$$= \frac{1(3)}{16(3x-1)} - \frac{1}{16(x+5)} = \frac{3(x+5)}{16(3x-1)(x+5)} - \frac{1(3x-1)}{16(x+5)(3x-1)}$$
$$= \frac{3x + 15 - (3x-1)}{16(3x-1)(x+5)} = \frac{3x + 15 - 3x + 1}{16(3x^2 + 15x - x - 5)} = \frac{16}{16(3x^2 + 14x - 5)} = \frac{1}{3x^2 + 14x - 5}$$

Chapter 6 – Partial Fractions

Example 2. Perform the definite integral below.

$$\int_{x=1}^{3} \frac{51x + 18}{8x^3 + 30x^2 + 18x} dx$$

Solution: First, note that we can factor x out of the denominator: $x(8x^2 + 30x + 18)$. Now factor[16] the remaining quadratic: $8x^2 + 30x + 18 = (4x + 3)(2x + 6)$. Putting all of this together, $8x^3 + 30x^2 + 18x = x(4x + 3)(2x + 6)$. There are 3 factors in all.

$$\frac{51x + 18}{8x^3 + 30x^2 + 18x} = \frac{a}{x} + \frac{b}{4x + 3} + \frac{e}{2x + 6}$$

Think[17] about making a common denominator of $x(4x + 3)(2x + 6)$. What would the new numerators be? These need to add up to the original numerator.

$$51x + 18 = a(4x + 3)(2x + 6) + bx(2x + 6) + ex(4x + 3)$$

First we'll distribute, and then we'll factor by powers of x.

$$51x + 18 = a(8x^2 + 24x + 6x + 18) + 2bx^2 + 6bx + 4ex^2 + 3ex$$

$$51x + 18 = 8ax^2 + 30ax + 18a + 2bx^2 + 6bx + 4ex^2 + 3ex$$

$$51x + 18 = (8a + 2b + 4e)x^2 + (30a + 6b + 3e)x + 18a$$

The left-hand side doesn't have any x^2 terms, so $8a + 2b + 4e = 0$. The coefficient of x on both sides tells us that $51 = 30a + 6b + 3e$. Equating the constant terms on both sides, we get $18 = 18a$. Solve this system of three equations.

$$8a + 2b + 4e = 0 \quad , \quad 51 = 30a + 6b + 3e \quad , \quad 18 = 18a$$

It's easy to solve for a in the right equation: $a = 1$. Plug this into the other equations.

$$8 + 2b + 4e = 0 \quad , \quad 51 = 30 + 6b + 3e$$

$$8 + 2b + 4e = 0 \quad , \quad 21 = 6b + 3e$$

$$4 + b + 2e = 0 \quad , \quad 7 = 2b + e$$

Isolate e in the right equation.

$$e = 7 - 2b$$

Plug this into the left equation.

---

[16] With the reasoning of Footnote 13, the factors of 8 could be 1 and 8 or 2 and 4 while the factors of 18 could be 18 and 1, 2 and 9, or 3 and 6. There are more combinations to try in this example (compared to the first example), like $(x + 3)(8x + 6)$ or $(4x + 2)(2x + 9)$. It turns out to be $(4x + 3)(2x + 6)$.

[17] In case you want to see it, you could do it yourself, but here it is: $\frac{a(4x+3)(2x+6)}{x(4x+3)(2x+6)} + \frac{bx(2x+6)}{x(4x+3)(2x+6)} + \frac{ex(4x+3)}{x(4x+3)(2x+6)}$. We only need the numerators to solve for the constants. (All the denominators are the same.)

$$4 + b + 2(7 - 2b) = 0$$
$$4 + b + 14 - 4b = 0$$
$$18 - 3b = 0 \quad \rightarrow \quad 18 = 3b \quad \rightarrow \quad \frac{18}{3} = 6 = b$$

Plug this into the equation $e = 7 - 2b$ from earlier: $e = 7 - 2(6) = 7 - 12 = -5$. Plug $a = 1, b = 6,$ and $c = -5$ into the partial fraction decomposition of the integral.

$$\int_{x=1}^{3} \frac{51x + 18}{8x^3 + 30x^2 + 18x} dx = \int_{x=1}^{3} \frac{dx}{x} + 6\int_{x=1}^{3} \frac{dx}{4x + 3} - 5\int_{x=1}^{3} \frac{dx}{2x + 6}$$

Let $u = 4x + 3, du = 4dx, \frac{du}{4} = dx, t = 2x + 6, dt = 2dx,$ and $\frac{dt}{2} = dx$. The new limits of u will be $4(1) + 3 = 4 + 3 = 7$ and $4(3) + 3 = 12 + 3 = 15$, while the new limits of t will be $2(1) + 6 = 2 + 6 = 8$ and $2(3) + 6 = 6 + 6 = 12$.

$$\int_{x=1}^{3} \frac{dx}{x} + 6\int_{x=1}^{3} \frac{dx}{4x + 3} - 5\int_{x=1}^{3} \frac{dx}{2x + 6} = \int_{x=1}^{3} \frac{dx}{x} + 6\int_{u=7}^{15} \frac{du}{4u} - 5\int_{t=8}^{12} \frac{dt}{2t}$$

$$= \int_{x=1}^{3} \frac{dx}{x} + \frac{3}{2}\int_{u=7}^{15} \frac{du}{u} - \frac{5}{2}\int_{t=8}^{12} \frac{dt}{t} = [\ln|x|]_{x=1}^{3} + \frac{3}{2}[\ln|u|]_{u=7}^{15} - \frac{5}{2}[\ln|t|]_{t=8}^{12}$$

$$= \ln 3 - \ln 1 + \frac{3}{2}\ln 15 - \frac{3}{2}\ln 7 - \frac{5}{2}\ln 12 + \frac{5}{2}\ln 8 = \ln\frac{3}{1} + \frac{3}{2}\ln\frac{15}{7} + \frac{5}{2}\ln\frac{8}{12}$$

$$= \ln 3 + \frac{3}{2}\ln\frac{15}{7} + \frac{5}{2}\ln\frac{2}{3} \approx 1.228$$

Alternate answer: $\frac{1}{2}\ln 9 + \frac{1}{2}\ln\left(\frac{15^3}{7^3}\right) + \frac{1}{2}\ln\left(\frac{2^5}{3^5}\right) = \frac{1}{2}\ln\left[\frac{(9)(3375)(32)}{(343)(243)}\right] = \frac{1}{2}\ln\left(\frac{4000}{343}\right)$.

Check the antiderivative: $f = \ln x + \frac{3}{2}\ln(4x + 3) - \frac{5}{2}\ln(2x + 6)$. Use the chain rule.

$$\frac{d}{dx}\left[\ln x + \frac{3}{2}\ln(4x + 3) - \frac{5}{2}\ln(2x + 6)\right] = \frac{1}{x} + \frac{3}{2(4x + 3)}\frac{d}{dx}(4x + 3) - \frac{5}{2(2x + 6)}\frac{d}{dx}(2x + 6)$$

$$= \frac{1}{x} + \frac{3(4)}{2(4x + 3)} - \frac{5(2)}{2(2x + 6)} = \frac{1}{x} + \frac{6}{4x + 3} - \frac{5}{2x + 6}$$

$$= \frac{1(4x + 3)(2x + 6)}{x(4x + 3)(2x + 6)} + \frac{6x(2x + 6)}{x(4x + 3)(2x + 6)} - \frac{5x(4x + 3)}{x(4x + 3)(2x + 6)}$$

$$= \frac{8x^2 + 24x + 6x + 18 + 12x^2 + 36x - (20x^2 + 15x)}{x(8x^2 + 24x + 6x + 18)}$$

$$= \frac{20x^2 + 66x + 18 - 20x^2 - 15x}{x(8x^2 + 30x + 18)} = \frac{51x + 18}{8x^3 + 30x^2 + 18x}$$

Example 3. Find the solution to the indefinite integral below.
$$\int \frac{x^3 + 5x^2 + 2x - 11}{x + 3} dx$$

Solution: This integral is different from Examples 1-2 in one very important regard; the degree (meaning highest power) of the denominator isn't greater than the degree of the numerator (since x isn't a higher power than x^3). In this case, first we need to divide the polynomial in the numerator by the polynomial in the denominator. If you know how to carry out the polynomial long division of $x^3 + 5x^2 + 2x - 11$ divided by $x + 3$ (which has a remainder), you can do that. If you're not fluent with the long division of polynomials, we'll present an alternate method here. Think about the polynomial you would need to multiply $x + 3$ by in order to obtain the polynomial $x^3 + 5x^2 + 2x - 11$. The polynomial would need to have the form $x^2 + ax + b + \frac{c}{x+3}$. The last constant, c, will turn out to be nonzero if the division has a remainder. (We'll see that it does have a remainder; that is $c \neq 0$ in this example.) Our long division problem says that $\frac{x^3+5x^2+2x-11}{x+3} = x^2 + ax + b + \frac{c}{x+3}$. We can rewrite this long division problem as a multiplication problem as follows:

$$(x + 3)\left(x^2 + ax + b + \frac{c}{x + 3}\right) = x^3 + 5x^2 + 2x - 11$$

Multiply out the left-hand side and reorganize it by powers of x.

$$x^3 + ax^2 + 3x^2 + bx + 3ax + 3b + c = x^3 + 5x^2 + 2x - 11$$
$$x^3 + (a + 3)x^2 + (b + 3a)x + 3b + c = x^3 + 5x^2 + 2x - 11$$

For this equation to be true for any value of x (except, obviously, $x = -3$, which is a domain problem for the original integrand), the coefficients of each power must match on both sides of the equation. Equating coefficients of like powers, we get:

$$a + 3 = 5 \quad , \quad b + 3a = 2 \quad , \quad 3b + c = -11$$

The left equation tells us that $a = 5 - 3 = 2$. The middle equation then gives $b + 3(2) = 2$, such that $b = 2 - 6 = -4$. The right equation then gives $3(-4) + c = -11$, which gives $c = -11 + 12 = 1$. Plug $a = 2$, $b = -4$, and $c = 1$ into our expression for the result of the polynomial long division: $x^2 + ax + b + \frac{c}{x+3} = x^2 + 2x - 4 + \frac{1}{x+3}$. This is equivalent to the original integrand, allowing us to rewrite the original integral in the following form.

$$\int \frac{x^3 + 5x^2 + 2x - 11}{x + 3} dx = \int \left(x^2 + 2x - 4 + \frac{1}{x + 3}\right) dx$$

$$= \int x^2 \, dx + 2 \int x \, dx - 4 \int dx + \int \frac{dx}{x + 3} = \frac{x^3}{3} + x^2 - 4x + \ln|x + 3| + c$$

Check the answer: Take a derivative of the answer with respect to x.

$$\frac{d}{dx}\left[\frac{x^3}{3} + x^2 - 4x + \ln(x + 3) + c\right] = x^2 + 2x - 4 + \frac{1}{x + 3}\frac{d}{dx}(x + 3)$$

$$= x^2 + 2x - 4 + \frac{1(1)}{x + 3} = x^2 + 2x - 4 + \frac{1}{x + 3}$$

To check that this agrees with the original integrand, multiply it by $x + 3$ and compare it with the original numerator.

$$(x + 3)\left(x^2 + 2x - 4 + \frac{1}{x + 3}\right) = x^3 + 2x^2 - 4x + 3x^2 + 6x - 12 + 1$$

$$= x^3 + 5x^2 + 2x - 11$$

Chapter 6 Problems

Directions: Use the method of partial fractions to perform each integral.

① $\displaystyle\int \frac{dt}{4t^2 - 25}$ ($|t| > 2.5$)

❷ $\displaystyle\int \frac{dy}{6y^2 - 7y - 20}$

❸ $\displaystyle\int \frac{14x+3}{4x^2+21x-18}\,dx$

4. $\displaystyle\int \frac{7z^2 + 29z - 36}{z^3 + z^2 - 6z}\,dz$

❺ $\int \dfrac{e^w\,dw}{e^{2w}-9}$

6 $\displaystyle\int \frac{x^3 - 5x^2 + 11x - 12}{x - 2}\,dx$

7 $\displaystyle\int_{\theta=\pi/6}^{\pi/3} \frac{\sin\theta}{4\cos\theta - \cos^2\theta}\,d\theta$

⑧ $\displaystyle\int_{x=0}^{3} \frac{x^2}{x+3}\,dx$

7 Integration by Parts

The strategy in this chapter is highly versatile as it applies to a great variety of integrals. The underlying idea stems from applying the **product rule** (of differentiation) to two functions, $f(x)$ and $g(x)$, which are multiplying one another:

$$\frac{d}{dx}f(x)g(x) = f(x)g'(x) + g(x)f'(x)$$

Here the prime (') indicates a derivative: $f'(x) = \frac{df}{dx}$ and $g'(x) = \frac{dg}{dx}$. If we integrate both sides of the above equation, we get

$$f(x)g(x) = \int f(x)g'(x)\,dx + \int g(x)f'(x)\,dx$$

If we use algebra to isolate the first integral, this becomes

$$\int f(x)g'(x)\,dx = f(x)g(x) - \int g(x)f'(x)\,dx$$

Let $u = f(x)$ and $v = g(x)$, such that $du = f(x)dx$ and $dv = g(x)dx$. Then we can express the above equation in the form

$$\int u\,dv = uv - \int v\,du$$

The equation above is known as **integration by parts**. All the math above is just to try to show you where the equation comes from; it comes from writing out the product rule and integrating both sides. What we'll focus on in this chapter is how to use the equation to integrate by parts. The main idea is to identify u and dv to make it easier to perform an integral. See the tips and examples that follow.

Note that if you use integration by parts with a **definite integral**, you need to evaluate the product of the functions, uv, over the limits (much like you do with the result of any definite integral). The limits i and f in the equation below represent the 'initial' and 'final' values of x. See the examples with definite integrals.

$$\int_i^f u\,dv = [uv]_i^f - \int_i^f v\,du$$

Tips for integrating by parts:

- You want to make two substitutions: u is one function **already in the integrand**, and dv is the **derivative** of another function (where udv put together form the given integrand). The goal is that vdu will be easier to integrate than udv. The bullet points below may help you identify u and dv in the given integrand.

- Take a **derivative** of u and find the **antiderivative** of dv. For example, consider $\int x^2 \ln x \, dx$. If you let $u = \ln x$, the rest of the integrand must be $dv = x^2 dx$, such that $\int u \, dv$ is the given integral. The new integral will have the form $uv - \int v \, du$. Since $u = \ln x$, a derivative tells us that $du = \frac{dx}{x}$. Since $dv = x^2 dx$, an antiderivative tells us that $v = \frac{x^3}{3}$. Plugging these expressions into the formula, we get $\frac{x^3}{3} \ln x - \int \frac{x^3}{3} \frac{1}{x} dx$, which is simpler than the given integral. See Example 1.

- **Don't forget the differential element** with du and dv. For example, in the bullet point above, $u = \ln x$ gives us $du = \frac{1}{x} dx$ (and not $du = \frac{1}{x}$), while $dv = x^2 dx$ includes dx (it isn't $dv = x^2$). It's common for students to focus on finding the derivative and antiderivative, and then to forget the differential element with du or dv. If you write $du =$ or $dv =$, remember the dx on the other side.

- Does the integrand contain any functions where its **derivative is simpler** to work with than the function itself? Two common examples are $\ln x$ and inverse trig functions, like $\sin^{-1} x$. For example, if $u = \ln x$ in the original integrand, the new integrand would include $du = \frac{dx}{x}$, which may be easier to work with. As another example, if $u = \sin^{-1} x$, then $du = \frac{1}{\sqrt{1-u^2}}$. See Examples 1-2.

- Does the integrand have an **extra x in the numerator**? For example, $\int x \cos x \, dx$ has an extra x. Let $u = x$. Then $dv = \cos x \, dx$, such that $udv = x \cos x \, dx$ makes the complete given integrand. Then $du = dx$ (from a derivative of $u = x$) and $v = \sin x$ (the antiderivative of $dv = \cos x \, dx$). The new integral will have the form $uv - \int v \, du = x \sin x - \int \sin x \, dx$. The new integral, $\int \sin x \, dx$, doesn't have an extra x in it like the original integral. This strategy uses the fact that the derivative of x with respect to x is the constant one; the new integral has a one where the original integral had an x. See Example 3.

Chapter 7 – Integration by Parts

- Integration by parts has the power of **reduction**. What does this mean? If an integral includes an exponent, like $\int x^2 \cos x\, dx$, integration by parts can often help to reduce the integral to a lower power. In this case, if you let $u = x^2$ and $dv = \cos x\, dx$, you get $du = 2x dx$ and $v = \sin x$, for which the formula gives $x^2 \sin x - \int 2x \sin x\, dx$. Observe that the new integral has x to the first power, which is reduced compared to the x^2 in the original integral. See Example 4.
- Sometimes, when you integrate by parts once, it doesn't seem like you have made any progress, yet persistence may still pay off. For example, an identity may make all the difference (as in Example 5), or it may help to **integrate by parts a second time** (see Example 6).

It will be handy to know the derivatives of the **inverse trig functions**:[18]

$$\frac{d}{dx}\sin^{-1} x = \frac{1}{\sqrt{1-x^2}} \quad , \quad \frac{d}{dx}\cos^{-1} x = \frac{-1}{\sqrt{1-x^2}}$$

$$\frac{d}{dx}\tan^{-1} x = \frac{1}{1+x^2} \quad , \quad \frac{d}{dx}\cot^{-1} x = \frac{-1}{1+x^2}$$

$$\frac{d}{dx}\sec^{-1} x = \frac{1}{|x|\sqrt{x^2-1}} \; (|x|>1) \quad , \quad \frac{d}{dx}\csc^{-1} x = \frac{-1}{|x|\sqrt{x^2-1}} \; (|x|>1)$$

Example 1. Find the solution to the indefinite integral below.

$$\int x^2 \ln x\, dx$$

Solution: Since the derivative of $\ln x$ is generally easier to work with than $\ln x$ itself, let $u = \ln x$. Then we need $dv = x^2 dx$ so that $u dv = x^2 \ln x\, dx$ makes the original integrand. Take a derivative of $u = \ln x$ to get $du = \frac{dx}{x}$, and find the antiderivative of $dv = x^2 dx$ to get $v = \frac{x^3}{3}$. Plug $u = \ln x$, $du = \frac{dx}{x}$, and $v = \frac{x^3}{3}$ into the formula for integration by parts, $\int u\, dv = uv - \int v\, du$.

$$\int x^2 \ln x\, dx = (\ln x)\left(\frac{x^3}{3}\right) - \int \frac{x^3}{3}\frac{1}{x}dx = \frac{x^3}{3}\ln x - \frac{1}{3}\int x^2\, dx$$

---

[18] If you're not in Quadrant I, you have to be careful. For example, this inverse sine formula is valid for $-\frac{\pi}{2} < \sin^{-1} x < \frac{\pi}{2}$ and this inverse cosine formula is valid for $0 < \cos^{-1} x < \pi$.

$$= \frac{x^3}{3}\ln x - \frac{1}{3}\frac{x^3}{3} + c = \frac{x^3}{3}\ln x - \frac{x^3}{9} + c = \frac{x^3}{3}\left(-\frac{1}{3} + \ln x\right) + c = \frac{x^3}{9}(-1 + 3\ln x) + c$$

Check the answer: Take a derivative with respect to x. Use the **product rule**.

$$\frac{d}{dx}\left(\frac{x^3}{3}\ln x - \frac{x^3}{9} + c\right) = \frac{x^3}{3}\frac{d}{dx}\ln x + \ln x\frac{d}{dx}\frac{x^3}{3} - \frac{3x^2}{9}$$

$$= \frac{x^3}{3}\left(\frac{1}{x}\right) + (\ln x)(x^2) - \frac{3x^2}{9} = \frac{x^2}{3} + x^2\ln x - \frac{x^2}{3} = x^2\ln x$$

Example 2. Perform the definite integral below.

$$\int_{t=0}^{1/2} \sin^{-1} t \, dt$$

Solution: Let $u = \sin^{-1} t$. Then dv is simply $dv = dt$ so that $u\,dv = \sin^{-1}t\,dt$ makes the original integrand. Take a derivative[19] of $u = \sin^{-1} t$ to get $du = \frac{dt}{\sqrt{1-t^2}}$, and find the antiderivative of $dv = dt$ to get $v = t$. Plug $u = \sin^{-1} t$, $du = \frac{dt}{\sqrt{1-t^2}}$, and $v = t$ into the formula for integration by parts, $\int u\,dv = uv - \int v\,du$.

$$\int_{t=0}^{1/2} \sin^{-1} t\,dt = [(\sin^{-1} t)(t)]_{t=0}^{1/2} - \int_{t=0}^{1/2} \frac{t}{\sqrt{1-t^2}}dt = [t\sin^{-1} t]_{t=0}^{1/2} - \int_{t=0}^{1/2} \frac{t}{\sqrt{1-t^2}}dt$$

Let $w = 1 - t^2$, such that $dw = -2t\,dt$, for which $-\frac{dw}{2} = t\,dt$. The new limits of integration will be from $1 - 0^2 = 1 - 0 = 1$ to $1 - \left(\frac{1}{2}\right)^2 = 1 - \frac{1}{4} = \frac{3}{4}$.

$$[t\sin^{-1} t]_{t=0}^{1/2} - \int_{t=0}^{1/2} \frac{t}{\sqrt{1-t^2}}dt = \frac{1}{2}\sin^{-1}\left(\frac{1}{2}\right) - 0\sin^{-1}(0) - \left(-\frac{1}{2}\right)\int_{w=1}^{3/4} \frac{dw}{\sqrt{w}}$$

$$= \frac{1}{2}\frac{\pi}{6} - 0 + \frac{1}{2}\int_{w=1}^{3/4} w^{-1/2}\,dw = \frac{\pi}{12} + \frac{1}{2}\left[\frac{w^{-\frac{1}{2}+1}}{-\frac{1}{2}+1}\right]_{w=1}^{3/4} = \frac{\pi}{12} + \frac{1}{2}\left[\frac{w^{1/2}}{1/2}\right]_{w=1}^{3/4}$$

$$= \frac{\pi}{12} + \frac{1}{2}[2w^{1/2}]_{w=1}^{3/4} = \frac{\pi}{12} + [\sqrt{w}]_{w=1}^{3/4} = \frac{\pi}{12} + \sqrt{\frac{3}{4}} - \sqrt{1} = \frac{\pi}{12} + \frac{\sqrt{3}}{2} - 1 \approx 0.128$$

---

[19] If you've forgotten your derivatives of inverse trig functions, see the text prior to Example 1.

Chapter 7 – Integration by Parts

Check the antiderivative: First, find the antiderivative in the definite integral above. Look for the expressions prior to plugging in the limits. The antiderivative is equal to $t\sin^{-1}t + \sqrt{w} = t\sin^{-1}t + \sqrt{1-t^2} = t\sin^{-1}t + (1-t^2)^{1/2}$. Take a derivative with respect to t. Use the **product rule** and the chain rule.

$$\frac{d}{dt}[t\sin^{-1}t + (1-t^2)^{1/2}] = t\frac{d}{dt}\sin^{-1}t + \sin^{-1}t\frac{d}{dt}t + \frac{1}{2}(1-t^2)^{-1/2}\frac{d}{dt}(1-t^2)$$

$$= t\frac{1}{\sqrt{1-t^2}} + (\sin^{-1}t)(1) + \frac{1}{2}\frac{1}{(1-t^2)^{1/2}}(-2t) = \frac{t}{\sqrt{1-t^2}} + \sin^{-1}t - \frac{t}{(1-t^2)^{1/2}}$$

$$= \frac{t}{\sqrt{1-t^2}} + \sin^{-1}t - \frac{t}{\sqrt{1-t^2}} = \sin^{-1}t$$

Example 3. Find the solution to the indefinite integral below.

$$\int x\cos x\, dx$$

Solution: Let $u = x$. Then we need $dv = \cos x\, dx$ so that $udv = x\cos x\, dx$ makes the original integrand. Take a derivative of $u = x$ to get $du = dx$, and find the antiderivative of $dv = \cos x\, dx$ to get $v = \sin x$. Plug $u = x$, $du = dx$, and $v = \sin x$ into the formula for integration by parts, $\int u\, dv = uv - \int v\, du$.

$$\int x\cos x\, dx = x\sin x - \int \sin x\, dx = x\sin x + \cos x + c$$

Check the answer: Take a derivative with respect to x. Use the **product rule**.

$$\frac{d}{dx}(x\sin x + \cos x + c) = x\frac{d}{dx}\sin x + \sin x\frac{d}{dx}x - \sin x$$

$$= x\cos x + (\sin x)(1) - \sin x = x\cos x + \sin x - \sin x = x\cos x$$

Example 4. Find the solution to the indefinite integral below.

$$\int x^2\cos x\, dx$$

Solution: Let $u = x^2$. Then we need $dv = \cos x\, dx$ so that $udv = x^2\cos x\, dx$ makes the original integrand. Take a derivative of $u = x^2$ to get $du = 2xdx$, and find the antiderivative of $dv = \cos x\, dx$ to get $v = \sin x$. Plug $u = x^2$, $du = 2xdx$, and $v = \sin x$ into the formula for integration by parts, $\int u\, dv = uv - \int v\, du$.

$$\int x^2\cos x\, dx = x^2\sin x - \int 2x\sin x\, dx$$

Techniques of Integration Calculus Practice Workbook

Wait a minute. Isn't the new integral supposed to be easier than the original integral? Well, it is; the new integral has x instead of x^2. If you read the tips for this chapter, one of them says not to be afraid to use integration by parts a second time, so we'll try that and see if it helps (and another tip says that integration by parts is great for reducing exponents; we already reduced x^2 to x, so maybe the second time will reduce x to a constant). Apply integration by parts to the new integral, $\int 2x \sin x \, dx$. Let $u_2 = 2x$. Then we need $dv_2 = \sin x \, dx$ so that $u_2 dv_2 = 2x \sin x \, dx$ makes the integrand of the second integral. Take a derivative of $u_2 = 2x$ to get $du_2 = 2dx$, and find the antiderivative of $dv_2 = \sin x \, dx$ to get $v_2 = -\cos x$. Plug $u_2 = 2x$, $du_2 = 2dx$, and $v_2 = -\cos x$ into the formula for integration by parts, $\int u_2 \, dv_2 = u_2 v_2 - \int v_2 \, du_2$.

$$\int 2x \sin x \, dx = 2x(-\cos x) - \int 2(-\cos x)\, dx = -2x \cos x + 2\int \cos x \, dx$$

$$= -2x \cos x + 2 \sin x + c$$

We're not finished yet. We only found that $\int 2x \sin x \, dx = -2x \cos x + 2 \sin x + c$. We need to plug this into $\int x^2 \cos x \, dx = x^2 \sin x - \int 2x \sin x \, dx$.

$$\int x^2 \cos x \, dx = x^2 \sin x - (-2x \cos x + 2 \sin x + c)$$

$$= x^2 \sin x + 2x \cos x - 2 \sin x - c = (x^2 - 2) \sin x + 2x \cos x - c$$

Note: Whether we write "+c" or "−c" doesn't really matter; adding a positive constant or subtracting a negative constant, for example, are equivalent. If you prefer, you can let $c_2 = -c$ and write $+c_2$. (Or you could name the first constant $-c$ so that the constant at the end can be called "+c.")

Check the answer: Take a derivative with respect to x. Use the **product rule**.

$$\frac{d}{dx}(x^2 \sin x + 2x \cos x - 2 \sin x + c)$$

$$= x^2 \frac{d}{dx} \sin x + \sin x \frac{d}{dx} x^2 + 2x \frac{d}{dx} \cos x + \cos x \frac{d}{dx}(2x) - 2 \cos x$$

$$= x^2 \cos x + (\sin x)(2x) + (2x)(-\sin x) + (\cos x)(2) - 2 \cos x$$

$$= x^2 \cos x + 2x \sin x - 2x \sin x + 2 \cos x - 2 \cos x = x^2 \cos x$$

Example 5. Perform the definite integral below.
$$\int_{\theta=0}^{\pi/3} \sec^3\theta \, d\theta$$

Solution: There are two ways to approach integration by parts when the integrand involves trig functions. One way is to try to identify u and dv in the given integrand, and another way is to first use trig identities. In this example, we'll first attempt to use the integrand as it's given (but if that doesn't work out, then we'll try to use trig identities). Let $u = \sec\theta$ and $dv = \sec^2\theta \, d\theta$ so that $u \, dv = \sec^3\theta \, d\theta$ makes the original integrand. Take a derivative of $u = \sec\theta$ to get $du = \sec\theta \tan\theta \, d\theta$, and find the antiderivative of $dv = \sec^2\theta \, d\theta$ to get $v = \tan\theta$. Plug $u = \sec\theta$, $du = \sec\theta \tan\theta \, d\theta$, and $v = \tan\theta$ into the formula for integration by parts, $\int u \, dv = uv - \int v \, du$.

$$\int_{\theta=0}^{\pi/3} \sec^3\theta \, d\theta = [\sec\theta \tan\theta]_{\theta=0}^{\pi/3} - \int_{\theta=0}^{\pi/3} \sec\theta \tan^2\theta \, d\theta$$

Is the new integral simpler than the original integral? It may not seem like it, but let's not give up on it too soon. Sometimes, with integration by parts, a little persistence pays off. Let's try using the trig identity $\sec^2\theta - 1 = \tan^2\theta$ (recall Chapter 3).

$$\int_{\theta=0}^{\pi/3} \sec^3\theta \, d\theta = [\sec\theta \tan\theta]_{\theta=0}^{\pi/3} - \int_{\theta=0}^{\pi/3} \sec\theta \, (\sec^2\theta - 1) \, d\theta$$

$$\int_{\theta=0}^{\pi/3} \sec^3\theta \, d\theta = [\sec\theta \tan\theta]_{\theta=0}^{\pi/3} - \int_{\theta=0}^{\pi/3} \sec^3\theta \, d\theta + \int_{\theta=0}^{\pi/3} \sec\theta \, d\theta$$

Oh, no! We got back the same integral that we started with. But wait! Don't erase the work and start over. A little algebra can save the day. Observe that the integral of $\sec^3\theta$ is positive on the left side, but negative on the right side. If we add the integral to both sides of the equation, we'll get twice the integral on the left side and this integral will cancel out on the right side (just as $x = c - x + y$ simplifies to $2x = c + y$).

$$2\int_{\theta=0}^{\pi/3} \sec^3\theta \, d\theta = [\sec\theta \tan\theta]_{\theta=0}^{\pi/3} + \int_{\theta=0}^{\pi/3} \sec\theta \, d\theta$$

Now, to solve for the integral of $\sec^3\theta$, we need to divide by 2 on both sides.

$$\int_{\theta=0}^{\pi/3} \sec^3\theta \, d\theta = \frac{1}{2}[\sec\theta\tan\theta]_{\theta=0}^{\pi/3} + \frac{1}{2}\int_{\theta=0}^{\pi/3} \sec\theta \, d\theta$$

$$= \frac{1}{2}\sec\frac{\pi}{3}\tan\frac{\pi}{3} - \frac{1}{2}\sec 0 \tan 0 + \frac{1}{2}[\ln|\sec\theta + \tan\theta|]_{\theta=0}^{\pi/3}$$

$$= \frac{1}{2}(2)(\sqrt{3}) - \frac{1}{2}(1)(0) + \frac{1}{2}\ln\left|\sec\frac{\pi}{3} + \tan\frac{\pi}{3}\right| - \frac{1}{2}\ln|\sec 0 + \tan 0|$$

$$= \sqrt{3} - 0 + \frac{1}{2}\ln(2+\sqrt{3}) - \frac{1}{2}\ln(1+0) = \sqrt{3} + \frac{1}{2}\ln(2+\sqrt{3}) - \frac{1}{2}\ln(1)$$

$$= \sqrt{3} + \frac{1}{2}\ln(2+\sqrt{3}) - 0 = \sqrt{3} + \frac{1}{2}\ln(2+\sqrt{3}) \approx 2.39$$

Alternate answer: It's fascinating that $\sqrt{3} + \frac{1}{2}\ln(2+\sqrt{3}) = \sqrt{3} - \frac{1}{2}\ln(2-\sqrt{3})$ are equivalent answers. It's because $\frac{1}{2-\sqrt{3}} = \frac{1}{2-\sqrt{3}}\frac{2+\sqrt{3}}{2+\sqrt{3}} = \frac{2+\sqrt{3}}{4+2\sqrt{3}-2\sqrt{3}-3} = \frac{2+\sqrt{3}}{1} = 2+\sqrt{3}$.

Check the antiderivative: Take a derivative with respect to θ. Use the **product rule** and the chain rule.

$$\frac{d}{d\theta}\left[\frac{1}{2}\sec\theta\tan\theta + \frac{1}{2}\ln(\sec\theta + \tan\theta)\right]$$

$$= \frac{1}{2}\tan\theta\frac{d}{d\theta}\sec\theta + \frac{1}{2}\sec\theta\frac{d}{d\theta}\tan\theta + \frac{1}{2(\sec\theta+\tan\theta)}\frac{d}{d\theta}(\sec\theta+\tan\theta)$$

$$= \frac{1}{2}\tan\theta\sec\theta\tan\theta + \frac{1}{2}\sec\theta\sec^2\theta + \frac{1}{2(\sec\theta+\tan\theta)}(\sec\theta\tan\theta+\sec^2\theta)$$

$$= \frac{1}{2}\tan^2\theta\sec\theta + \frac{1}{2}\sec^3\theta + \frac{\sec\theta(\tan\theta+\sec\theta)}{2(\sec\theta+\tan\theta)} = \frac{1}{2}(\sec^2\theta-1)\sec\theta + \frac{1}{2}\sec^3\theta + \frac{1}{2}\sec\theta$$

$$= \frac{1}{2}\sec^3\theta - \frac{1}{2}\sec\theta + \frac{1}{2}\sec^3\theta + \frac{1}{2}\sec\theta = \sec^3\theta$$

Example 6. Find the solution to the indefinite integral below.

$$\int e^{-x}\cos x \, dx$$

Solution: Let $u = e^{-x}$. Then we need $dv = \cos x \, dx$ so that $udv = e^{-x}\cos x \, dx$ makes the original integrand. Take a derivative of $u = e^{-x}$ to get $du = -e^{-x}dx$, and find the antiderivative of $dv = \cos x \, dx$ to get $v = \sin x$. Plug $u = e^{-x}$, $du = -e^{-x}dx$, and $v = \sin x$ into the formula for integration by parts, $\int u \, dv = uv - \int v \, du$.

$$\int e^{-x}\cos x \, dx = e^{-x}\sin x - \int \sin x \, (-e^{-x}) \, dx = e^{-x}\sin x + \int e^{-x}\sin x \, dx$$

Chapter 7 – Integration by Parts

We began with $\int e^{-x} \cos x \, dx$, and now we have $\int e^{-x} \sin x \, dx$. All we did so far is trade cosine for sine. How is that any better? It isn't any better yet, but if we don't give up, it will get better. Sometimes integrating by parts **a second time** is all you need. So let's try integrating the new integral, $\int e^{-x} \sin x \, dx$, by parts. Let $u_2 = e^{-x}$. Then we need $dv_2 = \sin x \, dx$ so that $u_2 dv_2 = e^{-x} \sin x \, dx$ makes the integrand of the second integral. Take a derivative of $u_2 = e^{-x}$ to get $du_2 = -e^{-x} dx$, and find the anti-derivative of $dv_2 = \sin x \, dx$ to get $v_2 = -\cos x$. Plug $u_2 = e^{-x}$, $du_2 = -e^{-x} dx$, and $v_2 = -\cos x$ into the formula for integration by parts, $\int u_2 \, dv_2 = u_2 v_2 - \int v_2 \, du_2$.

$$\int e^{-x} \sin x \, dx = e^{-x}(-\cos x) - \int (-\cos x)(-e^{-x}) \, dx$$

$$\int e^{-x} \sin x \, dx = -e^{-x} \cos x - \int e^{-x} \cos x \, dx$$

Be careful with the minus signs; the three minus signs make one minus sign. After integrating by parts a second time, we're back to the original integral (with cosine). How does that help? We can use the same trick that we learned in Example 5. First, substitute the above equation into $\int e^{-x} \cos x \, dx = e^{-x} \sin x + \int e^{-x} \sin x \, dx$.

$$\int e^{-x} \cos x \, dx = e^{-x} \sin x - e^{-x} \cos x - \int e^{-x} \cos x \, dx$$

Add $\int e^{-x} \cos x \, dx$ to both sides.

$$2 \int e^{-x} \cos x \, dx = e^{-x} \sin x - e^{-x} \cos x$$

Divide by 2. Since it's an indefinite integral, add a constant of integration.

$$\int e^{-x} \cos x \, dx = \frac{1}{2} e^{-x} \sin x - \frac{1}{2} e^{-x} \cos x + c = \frac{e^{-x}}{2}(\sin x - \cos x) + c$$

Check the answer: Take a derivative with respect to x. Use the **product rule**.

$$\frac{d}{dx}\left(\frac{1}{2} e^{-x} \sin x - \frac{1}{2} e^{-x} \cos x + c\right)$$

$$= \frac{1}{2} e^{-x} \frac{d}{dx} \sin x + \frac{1}{2} \sin x \frac{d}{dx} e^{-x} - \frac{1}{2} e^{-x} \frac{d}{dx} \cos x - \frac{1}{2} \cos x \frac{d}{dx} e^{-x}$$

$$= \frac{1}{2} e^{-x} \cos x + \frac{1}{2} \sin x (-e^{-x}) - \frac{1}{2} e^{-x}(-\sin x) - \frac{1}{2} \cos x (-e^{-x})$$

$$= \frac{1}{2} e^{-x} \cos x - \frac{1}{2} e^{-x} \sin x + \frac{1}{2} e^{-x} \sin x + \frac{1}{2} e^{-x} \cos x = e^{-x} \cos x$$

Chapter 7 Problems

Directions: Use the integration by parts formula to perform each integral.

❶ $\int \ln z \, dz$

❷ $\int \tan^{-1} y \, dy$ Note: This is inverse tangent (not 1 over tangent).

❸ $\displaystyle\int_{x=1}^{2} \sec^{-1} x \, dx$ Note: This is inverse secant (not 1 over secant).

❹ $\displaystyle\int w \cosh w \, dw$

5 $\int x \ln x \, dx$

6 $\int_{t=0}^{1} te^t \, dt$

❼ $\displaystyle\int y\tan^2 y\, dy$

❽ $\displaystyle\int \frac{\ln x}{x^3}\, dx$

⑨ $\displaystyle\int_{\varphi=0}^{\sqrt{\pi/2}} \varphi^3 \sin(\varphi^2)\, d\varphi$

⑩ $\displaystyle\int e^x \sin x\, dx$

⑪ $\displaystyle\int \frac{\ln y}{\sqrt{y}}\,dy$

⑫ $\displaystyle\int_{\theta=\pi/4}^{\pi/3} \sec^2\theta \ln(\tan\theta)\,d\theta$

⓭ $\int x^3 \sqrt{x^2 - 1}\, dx$

⓮ $\int \cos(\ln z)\, dz$

⑮ $\displaystyle\int_{w=0}^{1} \frac{w+1}{e^w}\,dw$

⑯ $\displaystyle\int \sin(\sqrt{\theta})\,d\theta$

17 $\int (\ln x)^2 \, dx$

18 $\displaystyle\int_{\theta=\pi/6}^{\pi/4} \cot^2 \theta \csc \theta \, d\theta$

19 $\int \sec^4 y \, dy$

8 Odd/even Functions over Symmetric Limits

A function is **odd** if $f(-x) = -f(x)$ for any value of x, and a function is **even** if $f(-x) = f(x)$ for any vale of x. Following are some examples.

- $f(x) = x^3$ is an odd function: $f(-x) = (-x)^3 = -x^3$. When $x = -2$, $f(-2) = (-2)^3 = -8$, whereas when $x = 2$, $f(2) = 2^3 = 8$. For any value of x, if you change the sign of x, it changes the sign of $f(x)$.
- $f(x) = \sin x$ is an odd function: $f(-x) = \sin(-x) = -\sin x$. When $x = -\frac{\pi}{6}$, $f\left(-\frac{\pi}{6}\right) = \sin\left(-\frac{\pi}{6}\right) = -\frac{1}{2}$, whereas when $x = \frac{\pi}{6}$, $f\left(\frac{\pi}{6}\right) = \sin\left(\frac{\pi}{6}\right) = \frac{1}{2}$.
- $f(x) = x^2$ is an even function: $f(-x) = (-x)^2 = x^2$. When $x = -3$, $f(-3) = (-3)^2 = 9$, and when $x = 3$, $f(3) = 3^2 = 9$. For any value of x, if you change the sign of x, it doesn't change the value (or the sign) of $f(x)$.
- $f(x) = \cos x$ is an even function: $f(-x) = \cos(-x) = \cos x$. When $x = -\frac{\pi}{6}$, $f\left(-\frac{\pi}{6}\right) = \cos\left(-\frac{\pi}{6}\right) = \frac{\sqrt{3}}{2}$, and when $x = \frac{\pi}{6}$, $f\left(\frac{\pi}{6}\right) = \cos\left(\frac{\pi}{6}\right) = \frac{\sqrt{3}}{2}$.
- You might wonder if all functions are either odd or even. They're not. Many functions are neither odd nor even. For example, consider $f = e^x$. When $x = -1$, $f(-1) = e^{-1} \approx 0.368$, whereas when $x = 1$, $f(1) = e^1 \approx 2.718$, so this function is neither odd nor even.

How does this relate to integrals? It does if an integral has **symmetric limits**, meaning that the lower limit is the negative of the upper limit, like the integral below, where the lower limit is $-a$ and the upper limit is a.

$$\int_{x=-a}^{a} f(x)\,dx$$

If $f(x)$ is an odd function in the integral above, the integral equals zero. You don't even have to bother doing the integration. Why does the integral of an odd function equal zero when the limits are symmetric? Because $f(-x) = -f(x)$, the area between $f(x)$ and the x-axis will have opposite sign from $x = -a$ to $x = 0$ as it has from $x = 0$ to $x = a$. For example, consider $f(x) = \sin x$ and $a = \frac{\pi}{6}$. In this case, $\sin x$ is negative

Chapter 8 – Odd/even Functions over Symmetric Limits

over the interval $-\frac{\pi}{6} \leq x \leq 0$ and $\sin x$ is positive over the interval $0 \leq x \leq \frac{\pi}{6}$. Not only that, but since $\sin(-x) = -\sin x$, the area above the curve (which is negative) from $-\frac{\pi}{6}$ to zero will be equal and opposite to the area below the curve (which is positive) from zero to $\frac{\pi}{6}$. Observe that

$$\int_{x=-\pi/6}^{\pi/6} \sin x \, dx = [-\cos x]_{x=-\pi/6}^{\pi/6} = -\cos\frac{\pi}{6} + \cos\left(-\frac{\pi}{6}\right) = -\frac{\sqrt{3}}{2} + \frac{\sqrt{3}}{2} = 0$$

More generally, if $f(x)$ is an **odd** function,

$$\int_{x=-a}^{a} f(x) \, dx = 0 \quad \text{if } f(-x) = -f(x)$$

If $f(x)$ is an **even** function,

$$\int_{x=-a}^{a} f(x) \, dx = 2 \int_{x=0}^{a} f(x) \, dx \quad \text{if } f(-x) = f(x)$$

For example, consider the integral of $\cos x$ from $-\frac{\pi}{6}$ to $-\frac{\pi}{6}$ and from $-\frac{\pi}{6}$ to zero:

$$\int_{x=-\pi/6}^{\pi/6} \cos x \, dx = [\sin x]_{x=-\pi/6}^{\pi/6} = \sin\frac{\pi}{6} - \sin\left(-\frac{\pi}{6}\right) = \frac{1}{2} - \left(-\frac{1}{2}\right) = \frac{1}{2} + \frac{1}{2} = 1$$

$$\int_{x=0}^{\pi/6} \cos x \, dx = [\sin x]_{x=0}^{\pi/6} = \sin\frac{\pi}{6} - \sin 0 = \frac{1}{2} - 0 = \frac{1}{2}$$

$$\int_{x=-\pi/6}^{\pi/6} \cos x \, dx = 2 \int_{x=0}^{\pi/6} \cos x \, dx$$

Make sure that the limits are **symmetric** (that the lower limit is the negative of the upper limit) before using the results above.

An odd function times and odd function makes an even function. For example, x^3 and $\sin x$ are each odd functions (all by themselves), but the combination $f(x) = x^3 \sin x$ is even. For example, when $x = -\frac{\pi}{2}$, $f\left(-\frac{\pi}{2}\right) = \left(-\frac{\pi}{2}\right)^3 \sin\left(-\frac{\pi}{2}\right) = -\frac{\pi^3}{8}(-1) = \frac{\pi^3}{8}$, and when $x = \frac{\pi}{2}$, $f\left(\frac{\pi}{2}\right) = \left(\frac{\pi}{2}\right)^3 \sin\left(\frac{\pi}{2}\right) = \frac{\pi^3}{8}(1) = \frac{\pi^3}{8}$.

Techniques of Integration Calculus Practice Workbook

An odd number of odd functions multiplied together make an odd function, whereas **an even number of odd functions multiplied together make an even function**. Any number of even functions multiplied together make an even function.

An odd function times an even function is odd. For example, $f(x) = \sin x \cos x$ is an odd function because $f(-x) = \sin(-x)\cos(-x) = -\sin x \cos x = -f(x)$.

A **polynomial** that only has odd powers (and no constant terms), like $x^5 + 4x^3 - 7x$, is odd. A polynomial that only has even powers or constant terms, like $x^4 + 5x^2 - 9$, is even. A polynomial that has both odd and even powers (or which has odd powers and a constant term, like $x^3 - 3$) is neither an odd function nor an even function.

If a function doesn't appear to be odd or even, meaning that it doesn't satisfy $f(-x) = -f(x)$ or $f(-x) = f(x)$, it may be possible to make a **substitution** so that after the substitution it is either odd or even. For example, consider $f(x) = \sin\left(x - \frac{\pi}{4}\right)$. Compare $f\left(-\frac{\pi}{4}\right) = \sin\left(-\frac{\pi}{4} - \frac{\pi}{4}\right) = \sin\left(-\frac{\pi}{2}\right) = -1$ with $f\left(\frac{\pi}{4}\right) = \sin\left(\frac{\pi}{4} - \frac{\pi}{4}\right) = \sin 0 = 0$. Here, $f\left(-\frac{\pi}{4}\right) = -1$ is negative, whereas $f\left(\frac{\pi}{4}\right) = 0$ is zero, so this function isn't odd or even in terms of x. However, if we define $u = x - \frac{\pi}{4}$, then $f(u) = \sin u$, and in terms of the new variable u, the function is odd. See Example 3.

Example 1. Perform the definite integral below.

$$\int_{\theta=-\pi/4}^{\pi/4} \tan\theta \, d\theta$$

Solution: Tangent is an odd function. For example, $\tan\left(-\frac{\pi}{4}\right) = -1$ has the opposite sign compared to $\tan\left(\frac{\pi}{4}\right) = 1$. For any value of θ, $\tan(-\theta) = -\tan\theta$. The integrand is an odd function and the limits are symmetric. Therefore, the integral is zero.

$$\int_{\theta=-\pi/4}^{\pi/4} \tan\theta \, d\theta = 0$$

Chapter 8 – Odd/even Functions over Symmetric Limits

Example 2. Perform the definite integral below.

$$\int_{x=-2}^{2} x^4 \, dx$$

Solution: x^4 is an even function. For example, $(-2)^4 = 16$ is equal to $2^4 = 16$. For any value of x, $(-x)^4 = x^4$. The integrand is an even function and the limits are symmetric. Therefore, we may change the lower limit to zero and double the integral.

$$\int_{x=-2}^{2} x^4 \, dx = 2 \int_{x=0}^{2} x^4 \, dx = 2 \left[\frac{x^5}{5} \right]_{x=0}^{2} = 2 \left(\frac{2^5}{5} - \frac{0^5}{5} \right) = \frac{2(2^5)}{5} = \frac{2^6}{5} = \frac{64}{5} = 12.8$$

Example 3. Perform the definite integral below.

$$\int_{x=-\pi/4}^{3\pi/4} \sin\left(x - \frac{\pi}{4}\right) dx$$

Solution: In terms of x, this integrand doesn't appear to be odd or even. For example, $\sin\left(-\frac{\pi}{4} - \frac{\pi}{4}\right) = \sin\left(-\frac{\pi}{2}\right) = -1$ whereas $\sin\left(\frac{\pi}{4} - \frac{\pi}{4}\right) = \sin 0 = 0$. Also, the limits in the above integral aren't symmetric. It looks like we'll have to do the integral the long way, so let's try the substitution $u = x - \frac{\pi}{4}$ and $du = dx$. When $x = -\frac{\pi}{4}$, $u = -\frac{\pi}{4} - \frac{\pi}{4} = -\frac{\pi}{2}$ and when $x = \frac{3\pi}{4}$, $u = \frac{3\pi}{4} - \frac{\pi}{4} = \frac{2\pi}{4} = \frac{\pi}{2}$. The integral becomes

$$\int_{u=-\pi/2}^{\pi/2} \sin u \, du = 0$$

Look at that! After making the substitution, the new integral is an odd function, since $\sin(-u) = -\sin u$, and the limits are symmetric. Therefore, the integral is zero.

Example 4. Perform the definite integral below.
$$\int_{x=-1}^{1} (x^4 + 2x^2 - 3) \sin x \, dx$$

Solution: The expression in parentheses is even, while sine is odd, and the product of an even function and an odd function is an odd function. For example, when $x = -0.5$, $[(-0.5)^4 + 2(-0.5)^2 - 3]$ is the same as $[(0.5)^4 + 2(0.5)^2 - 3]$, since x only has even powers, whereas $\sin(-0.5) = -\sin 0.5$. This means that $f(-0.5) = -f(0.5)$. This is true for any value of x. Since an odd function is being integrated over symmetric limits, the result is zero.

$$\int_{x=-1}^{1} (x^4 + 2x^2 - 3) \sin x \, dx = 0$$

Example 5. Perform the definite integral below.
$$\int_{x=-4}^{4} (x^5 - 6x^3 + 2x) \, dx$$

Solution: Since this polynomial has only odd powers of x and doesn't include any constant terms, it is an odd function. The integral is over symmetric limits. Therefore, this integral is zero.

$$\int_{x=-4}^{4} (x^5 - 6x^3 + 2x) \, dx = 0$$

Chapter 8 Problems

Directions: Use the odd or even nature of the integrand to evaluate each integral.

① $\displaystyle\int_{t=-7}^{7} 4t^5 \, dt$

② $\displaystyle\int_{y=-2}^{2} 5y^4 \, dy$

③ $\displaystyle\int_{x=-3}^{3} (x^7 - 4x^5 + 6x^3 - 9x) \, dx$

④ $\displaystyle\int_{\theta=-\pi/6}^{\pi/6} \cos(2\theta) \, d\theta$

❺ $\displaystyle\int_{\varphi=-\pi/3}^{\pi/3} \tan\varphi \, d\varphi$

❻ $\displaystyle\int_{x=-\pi/4}^{\pi/4} x \sin x \tan x \, dx$

❼ $\displaystyle\int_{z=-1}^{1} (z^3 - 2z)(z^5 + 3z^3)z \, dz$

❽ $\displaystyle\int_{x=-1}^{5} [(x-2)^3 + x - 2] \, dx$

9 Improper Integrals

There are two ways that a definite integral may be considered an **improper integral**: if the interval is infinite or if the function in the integrand grows infinite at one of the endpoints (or at some other point within the interval). Examples where the interval is infinite include $\int_{x=1}^{\infty} \frac{dx}{x^2}$, $\int_{x=-0}^{\infty} e^x \, dx$, and $\int_{x=-\infty}^{\infty} \frac{dx}{x^2+4}$. Examples where the integrand has a function that grows infinite include $\int_{x=0}^{1} \frac{dx}{x-1}$ (where $\frac{1}{x-1}$ grows infinite in the limit that x approaches 1) and $\int_{x=0}^{\pi/2} \tan x \, dx$ (where $\tan x$ grows infinite in the limit that x approaches $\frac{\pi}{2}$). Improper integrals are finite in some cases, but infinite in other cases. An improper integral that converges to a finite value is said to be **convergent**, whereas an improper integral that doesn't is said to be **divergent**.

Improper integrals have many important applications. Improper integrals are one way to test if an infinite series like $\sum_{n=1}^{\infty} \frac{1}{n^p}$ converges or diverges. Improper integrals appear in the context of random variables, probability distributions, and statistics (including their applications, such as quantum mechanics and statistical mechanics in physics), where the integrals naturally vary from $-\infty$ to $+\infty$. Improper integrals are also an inherent aspect of renormalization in the theory of collider physics.

The way to evaluate an improper integral is to apply the concept of a **limit**. If the integrand is finite for all x, but one of the limits is $+\infty$ or $-\infty$, then

$$\int_{x=a}^{\infty} f(x) \, dx = \lim_{t \to \infty} \int_{x=a}^{t} f(x) \, dx \quad , \quad \int_{x=-\infty}^{b} f(x) \, dx = \lim_{t \to -\infty} \int_{x=t}^{b} f(x) \, dx$$

If the integrand is finite for all x, but both limits are infinite, then split the integral into two separate integrals using any convenient finite real number c as follows.

$$\int_{x=-\infty}^{\infty} f(x) \, dx = \lim_{t \to -\infty} \int_{x=t}^{c} f(x) \, dx + \lim_{t \to \infty} \int_{x=c}^{t} f(x) \, dx$$

If the integrand grows infinite at one of the endpoints only (and not anywhere within the interval), then

$$\int_{x=a}^{b} f(x)\,dx = \lim_{t \to b^-} \int_{x=a}^{t} f(x)\,dx \quad , \quad \int_{x=a}^{b} f(x)\,dx = \lim_{t \to a^+} \int_{x=t}^{b} f(x)\,dx$$

The left case above is for a discontinuity at the upper limit while the right case above is for a discontinuity at the lower limit. If the integrand grows infinite in the interval (not at an endpoint), the integral may be divided into multiple integrals. For example, if the integrand is discontinuous only at c within the interval, then

$$\int_{x=a}^{b} f(x)\,dx = \int_{x=a}^{c} f(x)\,dx + \int_{x=c}^{b} f(x)\,dx$$

The formulas for splitting an improper integral up into two (or more) integrals work when each of the integrals is convergent. If any one of the integrals is divergent, then the overall integral is also divergent. We'll illustrate how to apply these formulas in the examples.

If the limit in one of the above formulas exists (and is finite) for a particular integral, the integral is said to be **convergent**. Otherwise, the integral is said to be **divergent**.

Example 1. Determine whether the improper integral is convergent or divergent. If it is convergent, find the value that it converges to.

$$\int_{x=1}^{\infty} \frac{dx}{x^2}$$

Solution: This integral is improper because the upper limit is infinite. (The integrand grows infinite as x approaches zero, but that's not an issue in this integral because 0 isn't included in the interval.) Use the formula for an infinite upper limit.

$$\int_{x=1}^{\infty} \frac{dx}{x^2} = \lim_{t \to \infty} \int_{x=1}^{t} \frac{dx}{x^2} = \lim_{t \to \infty} \int_{x=1}^{t} x^{-2}\,dx = \lim_{t \to \infty} \left[\frac{x^{-2+1}}{-2+1}\right]_{x=1}^{t} = \lim_{t \to \infty} \left[\frac{x^{-1}}{-1}\right]_{x=1}^{t}$$

$$= \lim_{t \to \infty} [-x^{-1}]_{x=1}^{t} = -\lim_{t \to \infty} \left[\frac{1}{x}\right]_{x=1}^{t} = -\lim_{t \to \infty} \left(\frac{1}{t} - \frac{1}{1}\right) = -(0 - 1) = -(-1) = 1$$

This integral is convergent and it converges to positive one.

Chapter 9 – Improper Integrals

Example 2. Determine whether the improper integral is convergent or divergent. If it is convergent, find the value that it converges to.

$$\int_{\theta=0}^{\pi/2} \tan\theta \, d\theta$$

Solution: This integral is improper because $\tan\theta$ grows infinite as θ approaches $\frac{\pi}{2}$. (This is easy to understand if you think of $\tan\theta$ as $\frac{\sin\theta}{\cos\theta}$. As θ approaches $\frac{\pi}{2}$, sine goes to one and cosine goes to zero. When sine is nearly one and cosine is barely greater than zero, the fraction is huge. Try taking the tangent of a number slightly less than 90 degrees, or $\frac{\pi}{2}$, on your calculator, being careful whether your calculator is in degrees or radians mode.) Use the formula for an integrand growing infinite at the upper limit.

$$\int_{\theta=0}^{\pi/2} \tan\theta \, d\theta = \lim_{t\to\frac{\pi}{2}^-} \int_{\theta=0}^{t} \tan\theta \, d\theta = \lim_{t\to\frac{\pi}{2}^-} [\ln|\sec\theta|]_{\theta=0}^{t} = \lim_{t\to\frac{\pi}{2}^-} (\ln|\sec t| - \ln|\sec 0|)$$

$$= \lim_{t\to\frac{\pi}{2}^-} (\ln|\sec t| - \ln 1) = \infty$$

This limit is infinite because $\sec t = \frac{1}{\cos t}$ and since $\cos t$ approaches zero as t approaches $\frac{\pi}{2}$ from below. Since the limit doesn't converge to a finite value, the integral diverges.

Example 3. Determine whether the improper integral is convergent or divergent. If it is convergent, find the value that it converges to.

$$\int_{x=-\infty}^{-1} \frac{dx}{x-2}$$

Solution: This integral is improper because the lower limit is $-\infty$. (The integrand grows infinite as x approaches 2, but that's not an issue in this integral because 2 isn't included in the interval.) Use the formula for a lower limit of $-\infty$.

$$\int_{x=-\infty}^{-1} \frac{dx}{x-2} = \lim_{t\to-\infty} \int_{x=t}^{-1} \frac{dx}{x-2}$$

Let $u = x - 2$ such that $du = dx$. The new limits of integration will be from $u = t - 2$ to $u = -1 - 2 = -3$ (found by plugging each limit into $u = x - 2$).

$$\lim_{t\to-\infty} \int_{x=t}^{-1} \frac{dx}{x-2} = \lim_{t\to-\infty} \int_{u=t-2}^{-3} \frac{du}{u} = \lim_{t\to-\infty} [\ln|u|]_{u=t-2}^{-3} = \lim_{t\to-\infty} [\ln|-3| - \ln|t-2|]_{u=t-2}^{-3}$$

$$= \ln 3 - \lim_{t\to-\infty} \ln|t-2| = -\infty$$

This limit is $-\infty$ because $\ln|t-2|$ grows infinite as t grows to $-\infty$. Since the limit doesn't converge to a finite value, the integral diverges.

Example 4. Determine whether the improper integral is convergent or divergent. If it is convergent, find the value that it converges to.

$$\int_{w=1}^{4} \frac{dw}{2-w}$$

Solution: Why is this integral improper? Neither limit is infinite. The integrand doesn't grow infinite at either limit. It's because the integrand grows infinite as w approaches 2 and because $w = 2$ is included in the interval (from 1 to 4). We need to split this integral into two parts joined at $w = 2$ as follows.

$$\int_{w=1}^{4} \frac{dw}{2-w} = \int_{x=1}^{2} \frac{dw}{2-w} + \int_{x=2}^{4} \frac{dw}{2-w}$$

<u>**The above equation is only valid if both of the integrals are convergent.**</u> So we'll analyze these integrals one at a time, and only join them together if both are convergent. In the left integral, the integrand grows infinite at the upper limit.

$$\int_{x=1}^{2} \frac{dw}{2-w} = \lim_{t\to 2^-} \int_{x=1}^{t} \frac{dw}{2-w}$$

Let $u = 2 - w$ such that $du = -dw$. Use $u = 2 - w$ to find the new limits: from $u = 2 - 1 = 1$ to $2 - t$.

$$\lim_{t\to 2^-} \int_{x=1}^{t} \frac{dw}{2-w} = -\lim_{t\to 2^-} \int_{u=1}^{2-t} \frac{du}{u} = -\lim_{t\to 2^-} [\ln|u|]_{u=1}^{2-t}$$

$$= -\lim_{t\to 2^-} (\ln|2-t| - \ln 1) = -\lim_{t\to 2^-} \ln|2-t| + 0 = \infty$$

This limit is infinite because $\ln(2-t)$ approaches $-\infty$ as t approaches 2 from below. (Try plugging in 1.99999 on your calculator.) Therefore, the integral diverges.

Chapter 9 – Improper Integrals

Important note: It would have been a **huge mistake** to evaluate $\int_{w=1}^{4} \frac{dw}{2-w}$ by using the antiderivative $-\ln|2-w|$ in the form $-[\ln|2-w|]_{w=1}^{4}$ because that would ignore the important point $w = 2$ where the integrand grows infinite. **When the integrand grows infinite within the interval, you must split it into separate integrals and verify that each integral is convergent before adding them together.**

Example 5. Determine whether the improper integral is convergent or divergent. If it is convergent, find the value that it converges to.

$$\int_{x=-\infty}^{\infty} \frac{dx}{x^2+9}$$

Solution: This integral is improper because both limits include ∞'s. (This integral doesn't grow infinite for any real value of x, since $x^2 + 9$ doesn't equal zero for any real value of x. Note that if the denominator had been $x + 9$ instead of $x^2 + 9$, then in that case the integrand would have grown infinite at -9, and if the denominator had been $x^2 - 9$ instead of $x^2 + 9$, the integrand would have grown infinite at -3 and 3. Set the denominator equal to zero in order to solve for any real values of x that make the integrand grow infinite.) Use the formula where both limits include ∞'s. Since the integrand remains finite for all real values of x, we may choose any real number for c; let's choose $c = 0$. (If the integrand had grown infinite for a real value of x, that's the value we would have used for c, just like the previous example; and if the integrand had grown infinite for multiple values of x, we would have split the integral into additional integrals in the same manner.)

$$\int_{x=-\infty}^{\infty} \frac{dx}{x^2+9} = \lim_{t \to -\infty} \int_{x=t}^{0} \frac{dx}{x^2+9} + \lim_{t \to \infty} \int_{x=0}^{t} \frac{dx}{x^2+9}$$

The above equation is only valid if both of the integrals are convergent. So we'll analyze these integrals one at a time, and only join them together if both are convergent. As we learned in Chapter 3, let $x = 3 \tan \theta$, for which $du = 3 \sec^2 \theta \, d\theta$. Solve for θ in $x = 3 \tan \theta$ to get $\theta = \tan^{-1}\left(\frac{x}{3}\right)$ and use this to find the new limits: from $\theta = \tan^{-1}\left(\frac{t}{3}\right)$ to $\tan^{-1} 0 = 0$ in the left integral and $\theta = \tan^{-1} 0 = 0$ to $\tan^{-1}\left(\frac{t}{3}\right)$ in the right integral. But we'll just do the left integral first and see if it's convergent before going further.

$$\lim_{t\to-\infty}\int_{x=t}^{0}\frac{dx}{x^2+9}=\lim_{t\to-\infty}\int_{\theta=\tan^{-1}(t/3)}^{0}\frac{3\sec^2\theta}{3^2\tan^2\theta+9}d\theta=\lim_{t\to-\infty}\int_{\theta=\tan^{-1}(t/3)}^{0}\frac{3\sec^2\theta}{9(1+\tan^2\theta)}d\theta$$

$$=\frac{1}{3}\lim_{t\to-\infty}\int_{\theta=\tan^{-1}(t/3)}^{0}\frac{\sec^2\theta}{\sec^2\theta}d\theta=\frac{1}{3}\lim_{t\to-\infty}\int_{\theta=\tan^{-1}(t/3)}^{0}d\theta=\frac{1}{3}\lim_{t\to-\infty}[\theta]_{\theta=\tan^{-1}(t/3)}^{0}$$

$$\lim_{t\to-\infty}\left[\frac{1}{3}(0)-\frac{1}{3}\tan^{-1}\left(\frac{t}{3}\right)\right]=0-\lim_{t\to-\infty}\frac{1}{3}\tan^{-1}\left(\frac{t}{3}\right)=0-\frac{1}{3}\left(-\frac{\pi}{2}\right)=\frac{\pi}{6}$$

When taking the inverse tangent of $\frac{t}{3}$ in the limit that t approaches $-\infty$, we're asking, "For which angle does the tangent of the angle grow to negative infinity?" The answer is that as the angle is approaching $-\frac{\pi}{2}$ radians from within Quadrant IV, the tangent of the angle grows very negative. Try taking the tangent of -89.99999 degrees on your calculator (making sure that it's in degrees mode). Since the above limit is finite, we'll move onto the second integral.

$$\lim_{t\to\infty}\int_{x=0}^{t}\frac{dx}{x^2+9}=\lim_{t\to\infty}\int_{\theta=0}^{\tan^{-1}(t/3)}\frac{3\sec^2\theta}{3^2\tan^2\theta+9}d\theta=\lim_{t\to\infty}\int_{\theta=0}^{\theta=\tan^{-1}(t/3)}\frac{3\sec^2\theta}{9(1+\tan^2\theta)}d\theta$$

$$=\frac{1}{3}\lim_{t\to\infty}\int_{\theta=0}^{\theta=\tan^{-1}(t/3)}\frac{\sec^2\theta}{\sec^2\theta}d\theta=\frac{1}{3}\lim_{t\to\infty}\int_{\theta=0}^{\theta=\tan^{-1}(t/3)}d\theta=\frac{1}{3}\lim_{t\to\infty}[\theta]_{\theta=0}^{\tan^{-1}(t/3)}$$

$$=\lim_{t\to\infty}\frac{1}{3}\tan^{-1}\left(\frac{t}{3}\right)-0=\frac{1}{3}\left(\frac{\pi}{2}\right)=\frac{\pi}{6}$$

Since both integrals converge to finite values, we may use the formula for adding them together as follows:

$$\int_{x=-\infty}^{\infty}\frac{dx}{x^2+9}=\lim_{t\to-\infty}\int_{x=t}^{0}\frac{dx}{x^2+9}+\lim_{t\to\infty}\int_{x=0}^{t}\frac{dx}{x^2+9}=\frac{\pi}{6}+\frac{\pi}{6}=\frac{2\pi}{6}=\frac{\pi}{3}$$

Chapter 9 – Improper Integrals

Example 6. Determine whether the improper integral is convergent or divergent. If it is convergent, find the value that it converges to.

$$\int_{z=0}^{1} \ln z \, dz$$

Solution: This integral is improper because $\ln z$ goes to $-\infty$ as z approaches zero from the right. Use the formula for an integrand going to $-\infty$ (or $+\infty$) at the lower limit.

$$\int_{z=0}^{1} \ln z \, dz = \lim_{t \to 0^+} \int_{z=t}^{1} \ln z \, dz$$

As we learned in Chapter 7, let $u = \ln z$ and $dv = dz$ such that $du = \frac{dz}{z}$ and $v = z$. Use the integration by parts formula $\int_i^f u \, dv = [uv]_i^f - \int_i^f v \, du$.

$$\lim_{t \to 0^+} \int_{z=t}^{1} \ln z \, dz = \lim_{t \to 0^+} [(\ln z)(z)]_{z=t}^{1} - \lim_{t \to 0^+} \int_{z=t}^{1} z \frac{dz}{z} = \lim_{t \to 0^+} [z \ln z]_{z=t}^{1} - \lim_{t \to 0^+} \int_{z=t}^{1} dz$$

$$= \lim_{t \to 0^+} (1 \ln 1 - t \ln t) - \lim_{t \to 0^+} [z]_{z=t}^{1} = 1 \ln 1 - \lim_{t \to 0^+} t \ln t - \lim_{t \to 0^+} (1 - t)$$

$$= 1(0) - \lim_{t \to 0^+} t \ln t - 1 + \lim_{t \to 0^+} t = 0 - \lim_{t \to 0^+} t \ln t - 1 + 0 = - \lim_{t \to 0^+} t \ln t - 1$$

To evaluate this limit, we need to apply **l'Hôpital's rule** (since t approaches zero while $\ln t$ goes to $-\infty$ as t approaches zero from the right; this is an indeterminate form). To apply l'Hôpital's rule (if you don't recall this valuable rule from first-semester calculus, you should take a moment to review it), note that $t = \frac{1}{1/t}$. Take a derivative of the numerator and denominator with respect to t and evaluate the ratio in the limit.

$$- \lim_{t \to 0^+} t \ln t - 1 = - \lim_{t \to 0^+} \frac{\ln t}{\frac{1}{t}} - 1 = - \frac{\frac{d}{dt} \ln t}{\frac{d}{dt} \frac{1}{t}} \bigg|_{t \to 0^+} - 1 = - \left(\frac{1/t}{-1/t^2} \bigg|_{t \to 0^+} \right) - 1$$

$$= - \left(\frac{1}{t} \div - \frac{1}{t^2} \right) \bigg|_{t \to 0^+} - 1 = - \left(\frac{1}{t} \times -t^2 \right) \bigg|_{t \to 0^+} - 1 = t |_{t \to 0^+} - 1 = 0 - 1 = -1$$

This integral is convergent and it converges to negative one. (It's negative because the function $\ln z$ lies below the z-axis in the interval $0 < z \le 1$.)

Chapter 9 Problems

Directions: Determine whether the improper integral is convergent or divergent. If it is convergent, find the value that it converges to.

❶ $\displaystyle\int_{x=1}^{\infty} \frac{dx}{\sqrt{x}}$

❷ $\displaystyle\int_{y=1}^{\infty} \frac{dy}{y^3}$

❸ $\displaystyle\int_{x=-1}^{1} \frac{dx}{\sqrt{x+1}}$

❹ $\displaystyle\int_{\theta=0}^{\pi/2} \sec^2\theta\, d\theta$

❺ $\displaystyle\int_{x=-\infty}^{\infty} \frac{x}{x^2+16}\, dx$

❻ $\displaystyle\int_{x=1}^{\infty} \frac{dx}{x\sqrt{x}}$

❼ $\displaystyle\int_{\varphi=0}^{\infty} \cos\varphi \, d\varphi$

❽ $\displaystyle\int_{z=0}^{1} z \ln z \, dz$

❾ $\displaystyle\int_{y=-3}^{3} \frac{dy}{y^2 - 4}$

⑩ $$\int_{x=-\infty}^{\infty} xe^{-x^2}\, dx$$

⑪ $$\int_{w=0}^{3} \frac{dw}{w^2 - 3w + 2}$$

⑫ $$\int_{x=1}^{\infty} \frac{dx}{x^p}$$ Note: p is a constant.

10 Double and Triple Integrals

A **double integral** or a **triple integral** can be performed one integral at a time as follows:
- Examine the limits of integration. **If any of the limits contains a variable, you must do that integral first.** For example, the upper limit for y in the integral below is x^2, so we would need to integrate over y before integrating over x.

$$\int_{x=0}^{2}\int_{y=0}^{x^2} xy^3\, dxdy$$

As another example, all of the limits in the integral below are constants, so we may do these integrals in either order.

$$\int_{x=0}^{1}\int_{y=0}^{\pi/2} x\sin y\, dxdy$$

- When you integrate over one variable, treat the other independent variables as if they were constants. We'll see this in the examples.
- After you find the antiderivative for one variable, evaluate it over the limits (as we always do with a definite integral) before performing the next integral. That is, if you integrate over x first, finish that integral (including evaluating it over the limits) before integrating over y.

Notation: In this book, we will first write $dxdy$ in the problem **regardless of which order we will eventually do the integrals in**, and we will write the limits for x on the left integral in the problem regardless of which order we will eventually do the integrals in. One reason is that we don't want you to 'cheat,' meaning we don't want you to look at the order of $dxdy$ in the problem or the order in which the limits are written in the problem to determine in which order to perform the integrals. Rather, **we want you to 'think' your way through the integrals by looking at the limits**. (Notes: For one, $dxdy = dydx$ is commutative multiplication because these are just scalars. For another, in applications of calculus, we work with differential area element dA and it's up to the student to setup the integral and decide whether $dA = dxdy$ or $dydx$. Some advanced texts would just write $d^2\vec{r}$ or d^2x for differential area.)

Chapter 10 – Double and Triple Integrals

Note: So that you won't have to guess which integration limits correspond to which variable, we'll clearly write "$x =$" or "$y =$" with each lower limit. (When textbooks don't do this, in those books the order in which $dxdy$ appears is important. But as we noted previously, more advanced texts just write dA, $d^2\vec{r}$, d^2x, leaving all the details up to the student. A good approach is to think your way through the process.)

Example 1. Evaluate the double integral below.

$$\int_{x=0}^{1}\int_{y=0}^{\pi/2} x \cos y\, dxdy$$

Solution: Since all of the lower and upper limits are constants, these integrals may be done in any order. We'll integrate over x first. Now we'll rearrange the integral so that it 'looks' like we're integrating over x first. Specifically, we'll put the x integral on the inside and the y integral on the outside. We may factor the $\cos y$ out of the x integral. Why? Because when we integrate over x, we treat the independent variable y as if it were a constant, and you can pull a constant out of an integral. (But don't pull $\cos y$ out of the y integral; y isn't constant when you're integrating over y.) We placed optional parentheses around the x integral. We're using them to help any students who are new to double integrals 'see' the x integral as separate from the y integral. See how everything with an x is inside parentheses, and everything with a y is outside of parentheses. This integrand is separable in that regard, unlike Example 3.

$$\int_{y=0}^{\pi/2} \cos y \left(\int_{x=0}^{1} x\, dx\right) dy$$

Now we'll integrate over x entirely, ignoring y until we finish. Since the answer to the definite integral with respect to x is just a number, we may pull this constant out of the y integral. (But this isn't always the case. Contrast this with Example 2). Then we'll carry out the y integration.

$$\int_{y=0}^{\pi/2} \cos y \left[\frac{x^2}{2}\right]_{x=0}^{1} dy = \int_{y=0}^{\pi/2} \cos y \left(\frac{1^2}{2} - \frac{0^2}{2}\right) dy = \int_{y=0}^{\pi/2} \frac{\cos y}{2} dy = \frac{1}{2}\int_{y=0}^{\pi/2} \cos y\, dy$$

$$= \frac{1}{2}[\sin y]_{y=0}^{\pi/2} = \frac{1}{2}\left(\sin\frac{\pi}{2} - \sin 0\right) = \frac{1}{2}(1 - 0) = \frac{1}{2} = 0.5$$

Techniques of Integration Calculus Practice Workbook

Example 2. Evaluate the double integral below.

$$\int_{x=0}^{2} \int_{y=0}^{x^2} xy^3 \, dxdy$$

Solution: The upper limit of the y integral, which is x^2, contains the variable x. Thus, we must do the y integral first. Similar to Example 1, we'll enclose the y integral in parentheses. To do this, we'll move the dy inside of the dx. When we integrate over y, we treat the independent variable x as if it were a constant, which allows us to factor x out of the y integral. Remember to evaluate the antiderivative of the y integral over the limits before doing the x integral. Be careful not to pull any x's outside of the x integral. (You may factor x out of the y integral, but not out of its own integral.)

$$\int_{x=0}^{2} x \left(\int_{y=0}^{x^2} y^3 \, dy \right) dx = \int_{x=0}^{2} x \left[\frac{y^4}{4} \right]_{y=0}^{x^2} dx = \int_{x=0}^{2} x \left[\frac{(x^2)^4}{4} - \frac{0^4}{4} \right] dx = \int_{x=0}^{2} x \left(\frac{x^8}{4} \right) dx$$

$$= \frac{1}{4} \int_{x=0}^{2} x^9 \, dx = \frac{1}{4} \left[\frac{x^{10}}{10} \right]_{x=0}^{2} = \frac{1}{40} (2^{10} - 0^{10}) = \frac{1024}{40} = \frac{128}{5} = 25.6$$

Example 3. Evaluate the double integral below.

$$\int_{x=-y}^{0} \int_{y=1}^{3} \frac{dxdy}{(2x+3y)^2}$$

Solution: The lower limit of the x integral, contains the variable y. Thus, we must do the x integral first. Similar to the previous examples, we'll enclose the x integral in parentheses. To do this, we'll move the integration symbol for x, including its limits, to the 'inside.' When we integrate over x, we treat the independent variable y as if it were a constant. Unlike the previous integral, there are no y's that can be factored out of the x integral; here, the y is 'stuck' with the $(2x + 3y)$; we just need to treat the y as we would any other constant (while we're doing the x integral).

$$\int_{y=1}^{3} \left(\int_{x=-y}^{0} \frac{dx}{(2x+3y)^2} \right) dy$$

Let $u = 2x + 3y$. Treating y as a constant (for the x integral), $du = 2dx$ or $\frac{du}{2} = dx$.

Use the equation $u = 2x + 3y$ along with the limits of the x integral (from $-y$ to 0) to find that the new limits for the x integral are from $u = 2(-y) + 3y = -2y + 3y = y$ to $u = 2(0) + 3y = 0 + 3y = 3y$.

$$\int_{y=1}^{3}\left(\int_{u=y}^{3y}\frac{1}{u^2}\frac{du}{2}\right)dy = \frac{1}{2}\int_{y=1}^{3}\left(\int_{u=y}^{3y}\frac{du}{u^2}\right)dy = \frac{1}{2}\int_{y=1}^{3}\left(\int_{u=y}^{3y}u^{-2}\,du\right)dy$$

$$= \frac{1}{2}\int_{y=1}^{3}\left[\frac{u^{-2+1}}{-2+1}\right]_{u=y}^{3y}dy = \frac{1}{2}\int_{y=1}^{3}\left[\frac{u^{-1}}{-1}\right]_{u=y}^{3y}dy = \frac{1}{2}\int_{y=1}^{3}\left[-\frac{1}{u}\right]_{u=y}^{3y}dy$$

$$= -\frac{1}{2}\int_{y=1}^{3}\left(\frac{1}{3y}-\frac{1}{y}\right)dy = -\frac{1}{2}\int_{y=1}^{3}\left(\frac{1}{3y}-\frac{3}{3y}\right)dy = -\frac{1}{2}\int_{y=1}^{3}\frac{-2}{3y}dy$$

$$= \int_{y=1}^{3}\frac{dy}{3y} = \frac{1}{3}[\ln|y|]_{y=0}^{3} = \frac{1}{3}(\ln 3 - \ln 1) = \frac{1}{3}(\ln 3 - 0) = \frac{\ln 3}{3} \approx 0.366$$

Example 4. Evaluate the triple integral below.

$$\int_{x=0}^{2}\int_{y=0}^{3}\int_{z=0}^{4}x^2 y\,dxdydz$$

Solution: Since all of the integration limits are constants, we may do these integrals in any order. We'll integrate over z first. When we integrate over z, we treat the other independent variables (x and y) as if they were constants.

$$\int_{x=0}^{2}\int_{y=0}^{3}\left(\int_{z=0}^{4}x^2 y\,dz\right)dxdy = \int_{x=0}^{2}\int_{y=0}^{3}x^2 y\left(\int_{z=0}^{4}dz\right)dxdy$$

$$= \int_{x=0}^{2}\int_{y=0}^{3}x^2 y[z]_{z=0}^{4}\,dxdy = \int_{x=0}^{2}\int_{y=0}^{3}x^2 y(4-0)\,dxdy = 4\int_{x=0}^{2}\int_{y=0}^{3}x^2 y\,dxdy$$

We'll integrate over y next, treating x as if it were a constant.

$$4\int_{x=0}^{2}\left(\int_{y=0}^{3}x^2 y\,dy\right)dx = 4\int_{x=0}^{2}x^2\left(\int_{y=0}^{3}y\,dy\right)dx = 4\int_{x=0}^{2}x^2\left[\frac{y^2}{2}\right]_{y=0}^{3}dx$$

$$= 4 \int_{x=0}^{2} x^2 \left[\frac{3^2}{2} - \frac{0^2}{2}\right]^3_{y=0} dx = 4 \int_{x=0}^{2} x^2 \left(\frac{9}{2}\right) dx = 18 \int_{x=0}^{2} x^2 \, dx$$

$$= 18 \left[\frac{x^3}{3}\right]^2_{x=0} = 6(2^3 - 0^3) = 6(8) = 48$$

Example 5. Evaluate the triple integral below.

$$\int_{x=0}^{y} \int_{y=1}^{2} \int_{z=0}^{x} \frac{x}{y} dx\, dy\, dz$$

Solution: The x and z integrals both have variable limits. Which integral should we do first? It's a bit of a logic problem:

- Since the x limits include y, we must integrate over x before y.
- Since the z limits include x, we must integrate over z before x.
- Since the y limits are constant, we'll save the y integral for last.

Therefore, we must do the z integral first (so that it will be done before x), the x integral second (so that it will be done before y), and the y integral last.

$$\int_{x=0}^{y} \int_{y=1}^{2} \left(\int_{z=0}^{x} \frac{x}{y} dz\right) dx\, dy = \int_{x=0}^{y} \int_{y=1}^{2} \frac{x}{y} \left(\int_{z=0}^{x} dz\right) dx\, dy = \int_{x=0}^{y} \int_{y=1}^{2} \frac{x}{y} [z]_{z=0}^{x} dx\, dy$$

$$= \int_{x=0}^{y} \int_{y=1}^{2} \frac{x}{y}(x-0)\, dx\, dy = \int_{x=0}^{y} \int_{y=1}^{2} \frac{x^2}{y} dx\, dy = \int_{y=1}^{2} \left(\int_{x=0}^{y} \frac{x^2}{y} dx\right) dy$$

$$= \int_{y=1}^{2} \frac{1}{y} \left(\int_{x=0}^{y} x^2\, dx\right) dy = \int_{y=1}^{2} \frac{1}{y} \left[\frac{x^3}{3}\right]_{x=0}^{y} dy = \int_{y=1}^{2} \frac{1}{y}\left(\frac{y^3}{3} - \frac{0^3}{3}\right) dy$$

$$= \frac{1}{3} \int_{y=1}^{2} y^2\, dy = \frac{1}{3}\left[\frac{y^3}{3}\right]_{y=1}^{2} = \frac{1}{9}(2^3 - 1^3) = \frac{7}{9} \approx 0.778$$

Chapter 10 Problems

Directions: Evaluate each integral.

❶ $\displaystyle\int_{x=0}^{\pi/3}\int_{y=0}^{1} e^{-y}\sec^2 x \, dx dy$

❷ $\displaystyle\int_{x=0}^{3}\int_{y=0}^{x} x^2 y^3 \, dx dy$

❸ $\displaystyle\int_{x=-y^2}^{y^2}\int_{y=1}^{2}\frac{x^2}{y^2}\,dx\,dy$

❹ $\displaystyle\int_{x=-1}^{3}\int_{y=1}^{x}(x^2-2xy+6)\,dx\,dy$

❺ $\displaystyle\int_{x=0}^{2}\int_{y=0}^{2x} (6x-3y)^3\, dx\, dy$

❻ $\displaystyle\int_{x=\pi/6}^{\pi/2}\int_{y=0}^{1} y\cos(xy)\, dx\, dy$

❼ $\displaystyle\int_{x=0}^{1}\int_{y=x}^{2x} e^{8x-3y}\,dxdy$

❽ $\displaystyle\int_{r=0}^{3}\int_{\theta=0}^{2\pi} r^2 \sin^2\theta\,drd\theta$

⑨ $\displaystyle\int_{x=1}^{y}\int_{y=1}^{4}\frac{y^2-x^2}{xy}dxdy$

⑩ $\displaystyle\int_{\theta=\pi/2}^{\pi}\int_{\varphi=0}^{2\pi}\sin\left(\frac{\theta}{3}-\frac{\varphi}{6}\right)d\theta d\varphi$

⑪ $$\int_{x=2}^{y} \int_{y=3}^{4} \frac{x-y+1}{xy-2x-y+2} dx dy$$

⑫ $$\int_{x=1}^{2} \int_{y=0}^{x} \frac{xy}{\sqrt{x^2+y^2}} dx dy$$

13 $$\int_{x=0}^{1}\int_{y=0}^{2}\int_{z=0}^{3} x^3 y^2 z \, dx\,dy\,dz$$

14 $$\int_{x=0}^{z}\int_{y=0}^{x}\int_{z=-1}^{1} (z^2 - xy) \, dx\,dy\,dz$$

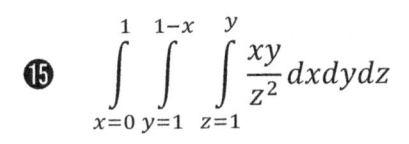

⑮ $\displaystyle\int_{x=0}^{1}\int_{y=1}^{1-x}\int_{z=1}^{y}\frac{xy}{z^2}dxdydz$

⑯ $\displaystyle\int_{x=-z}^{z}\int_{y=1}^{4}\int_{z=0}^{\sqrt{y}}\frac{dxdydz}{y}$

17 $$\int_{r=0}^{3}\int_{\varphi=0}^{\pi}\int_{\theta=0}^{2\pi} r^2 \sin\varphi \, dr d\varphi d\theta$$

18 $$\int_{r=0}^{3}\int_{\omega=1}^{2}\int_{t=0}^{\pi/6} r^2 \omega \cos(\omega t) \, dr d\omega dt$$

⑲ $\displaystyle\int_{x=0}^{y}\int_{y=0}^{1}\int_{z=0}^{x} xyze^{z^2}\,dxdydz$

Chapter 1 Solutions

❶ $\int 56x^7\, dx = \dfrac{56x^{7+1}}{7+1} + c = \dfrac{56x^8}{8} + c = 7x^8 + c$

Check the answer: $\dfrac{d}{dx}(7x^8 + c) = \dfrac{d}{dx}7x^8 + \dfrac{d}{dx}c = 7(8)x^{8-1} + 0 = 56x^7$

❷ $\int (t^3 - t)\, dt = \int t^3\, dt - \int t^1\, dt = \dfrac{t^{3+1}}{3+1} - \dfrac{t^{1+1}}{1+1} + c = \dfrac{t^4}{4} - \dfrac{t^2}{2} + c$

Notes: $t^1 = t$. There is no reason to have two separate constants of integration since the sum (or difference) of any two constants is just another constant. See Example 1.

Check the answer: $\dfrac{d}{dt}\left(\dfrac{t^4}{4} - \dfrac{t^2}{2} + c\right) = \dfrac{d}{dt}\dfrac{t^4}{4} - \dfrac{d}{dt}\dfrac{t^2}{2} + \dfrac{d}{dt}c = \dfrac{4t^3}{4} - \dfrac{2t^1}{2} + 0 = t^3 - t$

❸ $\displaystyle\int_{x=2}^{3} 12x^2\, dx = \left[\dfrac{12x^{2+1}}{2+1}\right]_{x=2}^{3} = \left[\dfrac{12x^3}{3}\right]_{x=2}^{3} = [4x^3]_{x=2}^{3} = 4(3)^3 - 4(2)^3$

$4(27) - 4(8) = 108 - 32 = 76$

Note: When performing a **definite** integral, there is no reason to include a constant of integration, since it would cancel out during the subtraction. In this problem, we would get $108 + c - (32 + c) = 108 + c - 32 - c = 108 - 32 = 76$.

Check the antiderivative: $\dfrac{d}{dx} 4x^3 = 4(3)x^{3-1} = 12x^2$

❹ $\displaystyle\int_{u=3}^{6} (u^2 - 4u + 2)\, du = \int_{u=3}^{6} u^2\, du - \int_{u=3}^{6} 4u^1\, du + \int_{u=3}^{6} 2\, du$

$= \left[\dfrac{u^{2+1}}{2+1}\right]_{u=3}^{6} - \left[\dfrac{4u^{1+1}}{1+1}\right]_{u=3}^{6} + [2u]_{u=3}^{6} = \left[\dfrac{u^3}{3}\right]_{u=3}^{6} - \left[\dfrac{4u^2}{2}\right]_{u=3}^{6} + [2u]_{u=3}^{6}$

$= \left[\dfrac{u^3}{3}\right]_{u=3}^{6} - [2u^2]_{u=3}^{6} + [2u]_{u=3}^{6} = \dfrac{6^3}{3} - \dfrac{3^3}{3} - 2(6)^2 - [-2(3)^2] + 2(6) - 2(3)$

$= \dfrac{216}{3} - \dfrac{27}{3} - 2(36) + 2(9) + 12 - 6 = 72 - 9 - 72 + 18 + 6 = 15$

Notes: Be careful with the sign of the middle integral. Since $[2u^2]_{u=3}^{6} = 2(6)^2 - 2(3)^2$, the overall minus sign makes this $-2(6)^2 + 2(3)^2$. The two terms have opposite signs.

Check the antiderivative: $\dfrac{d}{du}\left(\dfrac{u^3}{3} - 2u^2 + 2u\right) = \dfrac{d}{du}\dfrac{u^3}{3} - \dfrac{d}{du}2u^2 + \dfrac{d}{du}2u = u^2 - 4u + 2$

Note: Some calculators can actually give you numerical answers for definite integrals.

Techniques of Integration Calculus Practice Workbook

❺ $\int x^{2/3} \, dx = \int x^{2/3} \, dx = \dfrac{x^{\frac{2}{3}+1}}{\frac{2}{3}+1} + c = \dfrac{x^{5/3}}{5/3} + c = \dfrac{3}{5}x^{5/3} + c$

Notes: $\frac{2}{3} + 1 = \frac{2}{3} + \frac{3}{3} = \frac{5}{3}$ and $\frac{1}{5/3} = 1 \div \frac{5}{3} = 1 \times \frac{3}{5} = \frac{3}{5}$.

Check the answer: $\dfrac{d}{dx}\left(\dfrac{3}{5}x^{5/3} + c\right) = \dfrac{3}{5}\dfrac{5}{3}x^{5/3-1} + 0 = x^{2/3}$

❻ $\int \dfrac{dy}{\sqrt{y}} = \int \dfrac{dy}{y^{1/2}} = \int y^{-1/2} \, dy = \dfrac{y^{-\frac{1}{2}+1}}{-\frac{1}{2}+1} + c = \dfrac{y^{1/2}}{1/2} + c = 2y^{1/2} + c = 2\sqrt{y} + c$

Check the answer: $\dfrac{d}{dy}(2y^{1/2} + c) = y^{-1/2} + 0 = \dfrac{1}{y^{1/2}} = \dfrac{1}{\sqrt{y}}$

❼ $\int_{t=1}^{16} t^{3/4} \, dt = \left[\dfrac{t^{\frac{3}{4}+1}}{\frac{3}{4}+1}\right]_{t=1}^{16} = \left[\dfrac{t^{7/4}}{7/4}\right]_{t=1}^{16} = \dfrac{4}{7}[t^{7/4}]_{t=1}^{16} = \dfrac{4}{7}(16^{7/4} - 1^{7/4})$

$= \dfrac{4}{7}(128 - 1) = \dfrac{4(127)}{7} = \dfrac{508}{7}$

Notes: $\frac{3}{4} + 1 = \frac{3}{4} + \frac{4}{4} = \frac{7}{4}$ and $\frac{1}{7/4} = 1 \div \frac{7}{4} = 1 \times \frac{4}{7} = \frac{4}{7}$. You could find that $16^{7/4} = 128$ using a calculator. To do this by hand, first find the 4th root of 16. Which number raised to the fourth power equals 16? The answer is 2, since $2^4 = 16$. Now raise 2 to the 7th power to get 128.

Check the antiderivative: $\dfrac{d}{dt}\dfrac{4}{7}t^{7/4} = \dfrac{4}{7}\dfrac{7}{4}t^{3/4} = t^{3/4}$

❽ $\int_{z=3}^{24} \dfrac{dz}{z} = [\ln|z|]_{z=3}^{24} = \ln 24 - \ln 3 = \ln \dfrac{24}{3} = \ln 8 = \ln 2^3 = 3 \ln 2 \approx 2.07944$

Note: Recall the difference of logs formula $\ln y - \ln x = \ln \dfrac{y}{x}$ as well as $a \ln x = \ln x^a$.

Check the antiderivative: $\dfrac{d}{dz}\ln|z| = \dfrac{1}{z}$ (as in Example 3)

❾ $\int (\cos\theta - \sin\theta) \, d\theta = \int \cos\theta \, d\theta - \int \sin\theta \, d\theta = \sin\theta - (-\cos\theta) + c$

$= \sin\theta + \cos\theta + c$

Check the answer: $\dfrac{d}{d\theta}(\sin\theta + \cos\theta + c) = \dfrac{d}{d\theta}\sin\theta + \dfrac{d}{d\theta}\cos\theta + \dfrac{d}{d\theta}c = \cos\theta - \sin\theta$

❿ $\int (\cosh x - \sinh x) \, dx = \int \cosh x \, dx - \int \sinh x \, dx = \sinh x - \cosh x + c$

Answers with Full Solutions

Notes: sinh x and cosh x are different from the ordinary trig functions: sinh $x = \frac{e^x - e^{-x}}{2}$ and cosh $x = \frac{e^x + e^{-x}}{2}$. Their derivatives and antiderivatives don't follow the same sign pattern as the ordinary trig functions: $\frac{d}{dx}\cosh x = \sinh x$ and $\int \sinh x \, dx = \cosh x + c$ whereas $\frac{d}{dx}\cos x = -\sin x$ and $\int \sin x \, dx = -\cos x + c$.

Check the answer: $\frac{d}{dx}(\sinh x - \cosh x + c) = \cosh x - \sinh x$.

⑪ $\int_{\theta=\frac{\pi}{6}}^{\frac{\pi}{4}} \sin\theta \, d\theta = [-\cos\theta]_{\theta=\frac{\pi}{6}}^{\frac{\pi}{4}} = -[\cos\theta]_{\theta=\frac{\pi}{6}}^{\frac{\pi}{4}} = -\left(\cos\frac{\pi}{4} - \cos\frac{\pi}{6}\right)$

$= -\cos\frac{\pi}{4} + \cos\frac{\pi}{6} = -\frac{\sqrt{2}}{2} + \frac{\sqrt{3}}{2} = \frac{\sqrt{3} - \sqrt{2}}{2} \approx 0.1589186$

Notes: $\frac{\pi}{4}$ and $\frac{\pi}{6}$ radians correspond to 45° and 30°, respectively.

Check the antiderivative: $\frac{d}{d\theta}(-\cos\theta) = -\frac{d}{d\theta}\cos\theta = -(-\sin\theta) = \sin\theta$

⑫ $\int_{x=0}^{1} \frac{e^x - 1}{e^x} dx = \int_{x=0}^{1} 1 \, dx - \int_{x=0}^{1} e^{-x} \, dx = [x]_{x=0}^{1} - [-e^{-x}]_{x=0}^{1} = [x]_{x=0}^{1} + [e^{-x}]_{x=0}^{1}$

$= 1 - 0 + e^{-1} - e^{-0} = 1 + e^{-1} - 1 = e^{-1} = \frac{1}{e} \approx 0.3678794$

Notes: $\frac{e^x - 1}{e^x} = \frac{e^x}{e^x} - \frac{1}{e^x} = 1 - e^{-x}$ and $\int 1 \, dx = x + c$.

Check the antiderivative: $\frac{d}{dx}[x - (-e^{-x})] = \frac{d}{dx}(x + e^{-x}) = \frac{d}{dx}x + \frac{d}{dx}e^{-x} = 1 - e^{-x} =$
$1 - \frac{1}{e^x} = \frac{e^x}{e^x} - \frac{1}{e^x} = \frac{e^{-x} - 1}{e^x}$

⑬ $\int \sec^2\theta \, d\theta = \tan\theta + c$ Alternate answer: $\frac{\sin\theta}{\cos\theta} + c$

Check the answer: $\frac{d}{d\theta}(\tan\theta + c) = \frac{d}{d\theta}\tan\theta + \frac{d}{d\theta}c = \sec^2\theta + 0 = \sec^2\theta$

⑭ $\int \frac{\cos x}{\sin^2 x} dx = \int \frac{1}{\sin x} \frac{\cos x}{\sin x} dx = \int \csc x \cot x \, dx = -\csc x + c = -\frac{1}{\sin x} + c$

Notes: $\sin^2 x = \sin x \sin x$, $\csc x = \frac{1}{\sin x}$, and $\cot x = \frac{\cos x}{\sin x}$.

Check the answer: $\frac{d}{dx}(-\csc x + c) = -\frac{d}{dx}\csc x + \frac{d}{dx}c = -(-\csc x \cot x) + 0 =$
$\csc x \cot x = \frac{1}{\sin x} \frac{\cos x}{\sin x} = \frac{\cos x}{\sin^2 x}$

146

⑮ $\int \dfrac{d\varphi}{\sin^2 \varphi} = \int \csc^2 \varphi \, d\varphi = -\cot \varphi + c$

Alternate answers: $-\dfrac{\cos \varphi}{\sin \varphi} + c = -\dfrac{1}{\tan \varphi} + c$

Check the answer: $\dfrac{d}{d\varphi}(-\cot \varphi + c) = -\dfrac{d}{d\varphi}\cot \varphi + \dfrac{d}{d\varphi}c = -(-\csc^2 \varphi) + 0 = \csc^2 \varphi = \dfrac{1}{\sin^2 \varphi}$

Answers with Full Solutions

Chapter 2 Solutions

❶ $u = 3x^2 - 5, \dfrac{du}{dx} = 6x, du = 6xdx, \dfrac{du}{6} = xdx$

$\displaystyle\int \dfrac{x}{(3x^2 - 5)^3} dx = \int \dfrac{du}{6u^3} = \dfrac{1}{6}\int u^{-3}\, du = \dfrac{1}{6}\left(\dfrac{u^{-3+1}}{-3+1}\right) + c = \dfrac{1}{6}\left(\dfrac{u^{-2}}{-2}\right) + c$

$= -\dfrac{u^{-2}}{12} + c = -\dfrac{1}{12u^2} + c = -\dfrac{1}{12(3x^2 - 5)^2} + c$

Check the answer: $f = -\dfrac{1}{12u^2} + c = -\dfrac{1}{12(3x^2-5)^2} + c, u = 3x^2 - 5$

$\dfrac{df}{dx} = \dfrac{df}{du}\dfrac{du}{dx} = \dfrac{d}{du}\left(-\dfrac{1}{12u^2} + c\right)\dfrac{d}{dx}(3x^2 - 5) = -\dfrac{1}{12}(-2)\dfrac{1}{u^3}(6x) = \dfrac{x}{u^3} = \dfrac{x}{(3x^2 - 5)^3}$

Note: The **chain rule** is used to check the answers for this chapter, much like we used the chain rule to check the answers to the examples.

❷ $u = 9t + 4, \dfrac{du}{dt} = 9, du = 9dt, \dfrac{du}{9} = dt$

$\displaystyle\int \sqrt{9t + 4}\, dt = \int \dfrac{\sqrt{u}}{9}\, du = \dfrac{1}{9}\int u^{1/2}\, du = \dfrac{1}{9}\left(\dfrac{u^{\frac{1}{2}+1}}{\frac{1}{2}+1}\right) + c = \dfrac{1}{9}\left(\dfrac{u^{3/2}}{3/2}\right) + c$

$= \dfrac{1}{9}\dfrac{2}{3}u^{3/2} + c = \dfrac{2}{27}u^{3/2} + c = \dfrac{2}{27}(9t + 4)^{3/2} + c$

Alternate answer: $\dfrac{2}{27}(9t + 4)\sqrt{9t + 4} + c$ (since $x\sqrt{x} = x^1 x^{1/2} = x^{1+1/2} = x^{3/2}$)

Check the answer: $f = \dfrac{2}{27}u^{3/2} + c = \dfrac{2}{27}(9t + 4)^{3/2} + c, u = 9t + 4$

$\dfrac{df}{dt} = \dfrac{df}{du}\dfrac{du}{dt} = \dfrac{d}{du}\left(\dfrac{2}{27}u^{3/2} + c\right)\dfrac{d}{dt}(9t + 4) = \dfrac{2}{27}\dfrac{3}{2}u^{1/2}(9) = \dfrac{54}{54}u^{1/2} = \sqrt{u} = \sqrt{9t + 4}$

❸ $u = 2y - 1, \dfrac{du}{dy} = 2, du = 2dy, \dfrac{du}{2} = dy$

At $y = 1, u = 2(1) - 1 = 2 - 1 = 1$. At $y = 3, u = 2(3) - 1 = 6 - 1 = 5$.

$\displaystyle\int_{y=1}^{3}(2y - 1)^6\, dy = \int_{u=1}^{5}\dfrac{u^6}{2}\, du = \dfrac{1}{2}\int_{u=1}^{5} u^6\, du = \dfrac{1}{2}\left[\dfrac{u^7}{7}\right]_{u=1}^{5} = \dfrac{1}{14}[u^7]_{u=1}^{5}$

$= \dfrac{5^7 - 1^7}{14} = \dfrac{78{,}125 - 1}{14} = \dfrac{78{,}124}{14} = \dfrac{39{,}062}{7} \approx 5580.3$

Check the antiderivative: $f = \frac{u^7}{14} = \frac{(2y-1)^7}{14}, u = 2y - 1$

$\frac{df}{dy} = \frac{df}{du}\frac{du}{dy} = \frac{d}{du}\left(\frac{u^7}{14}\right)\frac{d}{dy}(2y-1) = \left(\frac{7u^6}{14}\right)(2) = u^6 = (2y-1)^6$

④ $u = x^4 + 2, \frac{du}{dx} = 4x^3, du = 4x^3 dx, \frac{du}{4} = x^3 dx$

At $x = 0, u = 0^4 + 2 = 2$. At $x = 1, u = 1^4 + 2 = 3$.

$\int_{x=0}^{1} x^3(x^4+2)^5 dx = \int_{u=2}^{3} \frac{u^5}{4} du = \frac{1}{4}\int_{u=2}^{3} u^5 du = \frac{1}{4}\left[\frac{u^6}{6}\right]_{u=2}^{3} = \frac{1}{24}[u^6]_{u=2}^{3}$

$= \frac{3^6 - 2^6}{24} = \frac{729 - 64}{24} = \frac{665}{24} \approx 27.71$

Check the antiderivative: $f = \frac{u^6}{24} = \frac{(x^4+2)^6}{24}, u = x^4 + 2$

$\frac{df}{dx} = \frac{df}{du}\frac{du}{dx} = \frac{d}{du}\left(\frac{u^6}{24}\right)\frac{d}{dx}(x^4 + 2) = \left(\frac{6u^5}{24}\right)(4x^3) = u^5 x^3 = x^3 u^5 = x^3(x^4+2)^5$

⑤ $u = 7t^3 - 4, \frac{du}{dt} = 21t^2, du = 21t^2 dt, \frac{du}{21} = t^2 dt$

$\int \frac{t^2}{7t^3 - 4} dt = \int \frac{du}{21u} = \frac{1}{21}\int \frac{du}{u} = \frac{\ln|u|}{21} + c = \frac{\ln|7t^3 - 4|}{21} + c$

Check the answer: $f = \frac{\ln|u|}{21} + c = \frac{\ln|7t^3-4|}{21} + c, u = 7t^3 - 4$

$\frac{df}{dt} = \frac{df}{du}\frac{du}{dt} = \frac{d}{du}\left(\frac{\ln|u|}{21} + c\right)\frac{d}{dt}(7t^3 - 4) = \left(\frac{1}{21u}\right)(21t^2) = \frac{t^2}{7t^3 - 4}$

⑥ $u = 3\theta, \frac{du}{d\theta} = 3, du = 3d\theta, \frac{du}{3} = d\theta$

$\int \cos(3\theta) d\theta = \int \frac{\cos u}{3} du = \frac{1}{3}\int \cos u\, du = \frac{\sin u}{3} + c = \frac{\sin(3\theta)}{3} + c$

Check the answer: $f = \frac{\sin u}{3} + c = \frac{\sin(3\theta)}{3} + c, u = 3\theta$

$\frac{df}{d\theta} = \frac{df}{du}\frac{du}{d\theta} = \frac{d}{du}\left(\frac{\sin u}{3} + c\right)\frac{d}{d\theta}(3\theta) = \left(\frac{\cos u}{3}\right)(3) = \cos(3\theta)$

⑦ $u = 9 - x^2, \frac{du}{dx} = -2x, du = -2x dx, -\frac{du}{2} = x dx$

At $x = 0, u = 9 - 0^2 = 9$. At $x = 2, u = 9 - 2^2 = 9 - 4 = 5$.

149

Answers with Full Solutions

$$\int_{x=0}^{2} \frac{x}{9-x^2} dx = -\int_{u=9}^{5} \frac{du}{2u} = -\frac{1}{2}\int_{u=9}^{5} \frac{du}{u} = -\frac{1}{2}[\ln|u|]_{u=9}^{5} = -\frac{1}{2}(\ln 5 - \ln 9)$$

$$= \frac{1}{2}(-\ln 5 + \ln 9) = \frac{1}{2}(\ln 9 - \ln 5) = \frac{1}{2}\ln\left(\frac{9}{5}\right) = \frac{\ln 1.8}{2} \approx 0.294$$

Note: We applied the rule $\ln p - \ln q = \ln \frac{p}{q}$.

Check the antiderivative: $f = -\frac{1}{2}\ln u = -\frac{1}{2}\ln(9 - x^2)$, $u = 9 - x^2$

$$\frac{df}{dx} = \frac{df}{du}\frac{du}{dx} = \frac{d}{du}\left(-\frac{\ln u}{2}\right)\frac{d}{dx}(9 - x^2) = \left(-\frac{1}{2u}\right)(-2x) = \frac{x}{u} = \frac{x}{9 - x^2}$$

❽ $u = 5t - 6, \frac{du}{dt} = 5, du = 5dt, \frac{du}{5} = dt$

At $t = 2$, $u = 5(2) - 6 = 10 - 6 = 4$. At $t = 3$, $u = 5(3) - 6 = 15 - 6 = 9$.

$$\int_{t=2}^{3} \frac{dt}{\sqrt{5t-6}} = \int_{t=4}^{9} \frac{du}{5\sqrt{u}} = \frac{1}{5}\int_{t=4}^{9} \frac{du}{u^{1/2}} = \frac{1}{5}\int_{t=4}^{9} u^{-1/2} du = \frac{1}{5}\left[\frac{u^{-\frac{1}{2}+1}}{-\frac{1}{2}+1}\right]_{u=4}^{9} = \frac{1}{5}\left[\frac{u^{1/2}}{1/2}\right]_{u=4}^{9}$$

$$= \frac{2}{5}[\sqrt{u}]_{u=4}^{9} = \frac{2}{5}(\sqrt{9} - \sqrt{4}) = \frac{2}{5}(3 - 2) = \frac{2}{5}(1) = \frac{2}{5} = 0.4$$

Check the antiderivative: $f = \frac{2}{5}\sqrt{u} = \frac{2}{5}\sqrt{5t-6}$, $u = 5t - 6$

$$\frac{df}{dt} = \frac{df}{du}\frac{du}{dt} = \frac{d}{du}\left(\frac{2}{5}\sqrt{u}\right)\frac{d}{dt}(5t - 6) = \left(\frac{2}{5}\frac{1}{2}\frac{1}{\sqrt{u}}\right)(5) = \frac{1}{\sqrt{5t-6}}$$

❾ $u = w^2, \frac{du}{dw} = 2w, du = 2wdw, \frac{du}{2} = wdw$

$$\int we^{w^2} dw = \frac{1}{2}\int e^u du = \frac{e^u}{2} + c = \frac{e^{w^2}}{2} + c$$

Check the answer: $f = \frac{e^u}{2} + c = \frac{e^{w^2}}{2} + c$, $u = w^2$

$$\frac{df}{dw} = \frac{df}{du}\frac{du}{dw} = \frac{d}{du}\left(\frac{e^u}{2} + c\right)\frac{d}{dw}(w^2) = \left(\frac{e^u}{2}\right)(2w) = we^{w^2}$$

❿ $u = x^3 + 2x, \frac{du}{dx} = 3x^2 + 2, du = (3x^2 + 2)dx$

$$\int (3x^2 + 2)\cos(x^3 + 2x) dx = \int \cos u \, du = \sin u + c = \sin(x^3 + 2x) + c$$

Check the answer: $f = \sin u + c = \sin(x^3 + 2x) + c$, $u = x^3 + 2x$

$$\frac{df}{dx} = \frac{df}{du}\frac{du}{dx} = \frac{d}{du}(\sin u + c)\frac{d}{dx}(x^3 + 2x) = (\cos u)(3x^2 + 2) = (3x^2 + 2)\cos(x^3 + 2x)$$

⑪ $u = \theta^3, \dfrac{du}{d\theta} = 3\theta^2, du = 3\theta^2 d\theta, \dfrac{du}{3} = \theta^2 d\theta$

At $\theta = 0, u = 0^3 = 0$. At $\theta = \sqrt[3]{\pi/4}, u = \left(\sqrt[3]{\pi/4}\right)^3 = \dfrac{\pi}{4}$.

$$\int_{\theta=0}^{\sqrt[3]{\pi/4}} \theta^2 \sin(\theta^3)\, d\theta = \int_{u=0}^{\pi/4} \dfrac{\sin u}{3}\, du = \dfrac{1}{3}\int_{u=0}^{\pi/4} \sin u\, du = \dfrac{1}{3}[-\cos u]_{u=0}^{\pi/4}$$

$$= \dfrac{1}{3}\left(-\cos\dfrac{\pi}{4} + \cos 0\right) = \dfrac{1}{3}\left(-\dfrac{\sqrt{2}}{2} + 1\right) = \dfrac{1}{3}\left(-\dfrac{\sqrt{2}}{2} + \dfrac{2}{2}\right) = \dfrac{1}{3}\left(\dfrac{-\sqrt{2}+2}{2}\right)$$

$$= \dfrac{2 - \sqrt{2}}{6} \approx 0.0976$$

Check the antiderivative: $f = -\dfrac{1}{3}\cos u = -\dfrac{1}{3}\cos(\theta^3), u = \theta^3$

$\dfrac{df}{d\theta} = \dfrac{df}{du}\dfrac{du}{d\theta} = \dfrac{d}{du}\left(-\dfrac{1}{3}\cos u\right)\dfrac{d}{d\theta}(\theta^3) = \left(-\dfrac{1}{3}\right)(-\sin u)(3\theta^2) = \theta^2 \sin(\theta^3)$

⑫ $u = \dfrac{1}{\varphi}, \dfrac{du}{d\varphi} = -\dfrac{1}{\varphi^2}, du = -\dfrac{d\varphi}{\varphi^2}, -du = \dfrac{d\varphi}{\varphi^2}$

At $\varphi = \dfrac{2}{\pi}, u = \left(\dfrac{2}{\pi}\right)^{-1} = \dfrac{\pi}{2}$. At $\varphi = \dfrac{6}{\pi}, u = \left(\dfrac{6}{\pi}\right)^{-1} = \dfrac{\pi}{6}$.

$$\int_{\varphi=2/\pi}^{6/\pi} \dfrac{1}{\varphi^2}\sin\left(\dfrac{1}{\varphi}\right) d\varphi = -\int_{u=\pi/2}^{\pi/6} \sin u\, du = -[-\cos u]_{u=\pi/2}^{\pi/6} = [\cos u]_{u=\pi/2}^{\pi/6}$$

$$= \cos\dfrac{\pi}{6} - \cos\dfrac{\pi}{2} = \dfrac{\sqrt{3}}{2} - 0 = \dfrac{\sqrt{3}}{2} \approx 0.866$$

Check the antiderivative: $f = \cos u = \cos\left(\dfrac{1}{\varphi}\right), u = \dfrac{1}{\varphi}$

$\dfrac{df}{d\varphi} = \dfrac{df}{du}\dfrac{du}{d\varphi} = \dfrac{d}{du}(\cos u)\dfrac{d}{d\varphi}\left(\dfrac{1}{\varphi}\right) = (-\sin u)\left(-\dfrac{1}{\varphi^2}\right) = \dfrac{1}{\varphi^2}\sin\left(\dfrac{1}{\varphi}\right)$

⑬ $u = \varphi\sqrt{\varphi} = \varphi^1\varphi^{1/2} = \varphi^{3/2}, \dfrac{du}{d\varphi} = \dfrac{3}{2}\varphi^{1/2} = \dfrac{3}{2}\sqrt{\varphi}, du = \dfrac{3}{2}\sqrt{\varphi}\,d\varphi, \dfrac{2du}{3} = \sqrt{\varphi}\,d\varphi$

$$\int \sqrt{\varphi}\sin(\varphi\sqrt{\varphi})\,d\varphi = \int \dfrac{2}{3}\sin u\, du = -\dfrac{2}{3}\cos u + c = -\dfrac{2}{3}\cos(\varphi\sqrt{\varphi}) + c$$

Check the answer: $f = -\dfrac{2}{3}\cos u + c = -\dfrac{2}{3}\cos(\varphi\sqrt{\varphi}) + c, u = \varphi\sqrt{\varphi} = \varphi^{3/2}$

$\dfrac{df}{d\varphi} = \dfrac{df}{du}\dfrac{du}{d\varphi} = \dfrac{d}{du}\left(-\dfrac{2}{3}\cos u + c\right)\dfrac{d}{d\varphi}(\varphi^{3/2}) = \left(\dfrac{2}{3}\sin u\right)\left(\dfrac{3}{2}\varphi^{1/2}\right) = \sqrt{\varphi}\sin(\varphi\sqrt{\varphi})$

Answers with Full Solutions

14 $u = \sqrt{t} - 1, \dfrac{du}{dt} = \dfrac{1}{2\sqrt{t}}, du = \dfrac{dt}{2\sqrt{t}}, 2du = \dfrac{dt}{\sqrt{t}}$

$\displaystyle\int \dfrac{(\sqrt{t}-1)^4}{\sqrt{t}} dt = \int 2u^4\, du = \dfrac{2u^5}{5} + c = \dfrac{2}{5}(\sqrt{t}-1)^5 + c$

Check the answer: $f = \dfrac{2u^5}{5} + c = \dfrac{2}{5}(\sqrt{t}-1)^5 + c, u = \sqrt{t} - 1$

$\dfrac{df}{dt} = \dfrac{df}{du}\dfrac{du}{dt} = \dfrac{d}{du}\left(\dfrac{2u^5}{5} + c\right)\dfrac{d}{dt}(\sqrt{t}-1) = (2u^4)\left(\dfrac{1}{2\sqrt{t}}\right) = \dfrac{(\sqrt{t}-1)^4}{\sqrt{t}}$

15 $(3z+2)^2 = (3z+2)(3z+2) = 9z^2 + 12z + 4$

$\displaystyle\int_{z=0}^{2} (9z^2 + 12z + 4)^4\, dz = \int_{z=0}^{2} [(3z+2)^2]^4\, dz = \int_{z=0}^{2} (3z+2)^8\, dz$

$u = 3z + 2, du = 3dz, \dfrac{du}{3} = dz$

At $z = 0, u = 3(0) + 2 = 2$. At $z = 2, u = 3(2) + 2 = 6 + 2 = 8$.

$\displaystyle\int_{z=0}^{2} (9z^2 + 12z + 4)^4\, dz = \int_{z=0}^{2}(3z+2)^8\, dz = \int_{u=2}^{8} \dfrac{u^8}{3}\, du = \dfrac{1}{3}\left[\dfrac{u^9}{9}\right]_{u=2}^{8} = \dfrac{[u^9]_{u=2}^{8}}{27}$

$= \dfrac{8^9 - 2^9}{27} = \dfrac{134{,}217{,}728 - 512}{27} = \dfrac{134{,}217{,}216}{27} = 4{,}971{,}008$

Why is the answer so huge? When $z = 2$, the function in the integrand equals $[9(2)^2 + 12(2) + 4]^4 = (36 + 24 + 4)^4 = 64^4 = 16{,}777{,}216$, so it seems reasonable for the area under the curve from $z = 0$ to $z = 2$ to have seven digits.

Check the antiderivative: $f = \dfrac{u^9}{27} = \dfrac{(3z+2)^9}{27}, u = 3z + 2$

$\dfrac{df}{dz} = \dfrac{df}{du}\dfrac{du}{dz} = \dfrac{d}{du}\left(\dfrac{u^9}{27}\right)\dfrac{d}{dz}(3z+2) = \left(\dfrac{u^8}{3}\right)(3) = u^8 = (3z+2)^8$

$= [(3z+2)^2]^4 = [(3z+2)(3z+2)]^4 = (9z^2 + 12z + 4)^4$

16 $u = 8\theta - \pi, \dfrac{du}{d\theta} = 8, du = 8d\theta, \dfrac{du}{8} = d\theta$

At $\theta = 0, u = 8(0) - \pi = -\pi$. At $\theta = \dfrac{\pi}{6}, u = 8\left(\dfrac{\pi}{6}\right) - \pi = \dfrac{4\pi}{3} - \dfrac{3\pi}{3} = \dfrac{\pi}{3}$.

$\displaystyle\int_{\theta=0}^{\pi/6} \cos(8\theta - \pi)\, d\theta = \int_{u=-\pi}^{\pi/3} \dfrac{\cos u}{8}\, du = \dfrac{1}{8}\int_{u=-\pi}^{\pi/3} \cos u\, du = \dfrac{1}{8}[\sin u]_{u=-\pi}^{\pi/3}$

$$= \frac{1}{8}\left[\sin\left(\frac{\pi}{3}\right) - \sin(-\pi)\right] = \frac{1}{8}\frac{\sqrt{3}}{2} - 0 = \frac{\sqrt{3}}{16} \approx 0.108$$

Check the antiderivative: $f = \frac{\sin u}{8} = \frac{\sin(8\theta - \pi)}{8}, u = 8\theta - \pi$

$$\frac{df}{d\theta} = \frac{df}{du}\frac{du}{d\theta} = \frac{d}{du}\left(\frac{\sin u}{8}\right)\frac{d}{d\theta}(8\theta - \pi) = \left(\frac{\cos u}{8}\right)(8) = \cos u = \cos(8\theta - \pi)$$

⑰ $\sqrt{\dfrac{x-1}{x^5}} = \sqrt{\dfrac{x-1}{x^4 x^1}} = \sqrt{\dfrac{1}{x^4}}\sqrt{\dfrac{x-1}{x}} = \dfrac{1}{x^2}\sqrt{\dfrac{x}{x} - \dfrac{1}{x}} = \dfrac{1}{x^2}\sqrt{1 - \dfrac{1}{x}}$

$$\int\sqrt{\frac{x-1}{x^5}}\,dx = \int\frac{1}{x^2}\sqrt{1 - \frac{1}{x}}\,dx$$

$$u = 1 - \frac{1}{x}, \frac{du}{dx} = -\left(-\frac{1}{x^2}\right) = \frac{1}{x^2}, du = \frac{dx}{x^2}$$

$$\int\sqrt{\frac{x-1}{x^5}}\,dx = \int\frac{1}{x^2}\sqrt{1-\frac{1}{x}}\,dx = \int\sqrt{u}\,du = \int u^{1/2}\,du = \frac{u^{\frac{1}{2}+1}}{\frac{1}{2}+1} + c = \frac{u^{3/2}}{3/2} + c$$

$$= \frac{2u^{3/2}}{3} + c = \frac{2}{3}\left(1 - \frac{1}{x}\right)^{3/2} + c$$

Alternate answer: $\frac{2}{3}\left(1 - \frac{1}{x}\right)\sqrt{1 - \frac{1}{x}} + c$

Check the answer: $f = \frac{2u^{3/2}}{3} + c = \frac{2}{3}\left(1 - \frac{1}{x}\right)^{3/2} + c, u = 1 - \frac{1}{x}$

$$\frac{df}{dx} = \frac{df}{du}\frac{du}{dx} = \frac{d}{du}\left(\frac{2u^{3/2}}{3}\right)\frac{d}{dx}\left(1 - \frac{1}{x}\right) = u^{1/2}(-1)\left(-\frac{1}{x^2}\right) = \frac{1}{x^2}u^{1/2} = \frac{1}{x^2}\sqrt{1 - \frac{1}{x}}$$

$$= \sqrt{\frac{1}{x^4}}\sqrt{1 - \frac{1}{x}} = \sqrt{\frac{1}{x^4}\left(1 - \frac{1}{x}\right)} = \sqrt{\frac{1}{x^4} - \frac{1}{x^5}} = \sqrt{\frac{x}{x^5} - \frac{1}{x^5}} = \sqrt{\frac{x-1}{x^5}}$$

⑱ $u = x^4 + 6x^2, \dfrac{du}{dx} = 4x^3 + 12x = 4(x^3 + 3x), du = 4(x^3 + 3x)dx, \dfrac{du}{4} = (x^3 + 3x)dx$

$$\int\frac{x^3 + 3x}{(x^4 + 6x^2)^2}\,dx = \int\frac{du}{4u^2} = \frac{1}{4}\int u^{-2}\,du = \frac{1}{4}\left(\frac{u^{-2+1}}{-2+1}\right) + c = \frac{1}{4}\left(\frac{u^{-1}}{-1}\right) + c$$

$$= -\frac{u^{-1}}{4} + c = -\frac{1}{4u} + c = -\frac{1}{4(x^4 + 6x^2)} + c = -\frac{1}{4x^4 + 24x^2} + c$$

153

Answers with Full Solutions

Check the answer: $f = -\frac{1}{4u} + c = -\frac{1}{4(x^4+6x^2)} + c, u = x^4 + 6x^2$

$\frac{df}{dx} = \frac{df}{du}\frac{du}{dx} = \frac{d}{du}\left(-\frac{1}{4u} + c\right)\frac{d}{dx}(x^4 + 6x^2) = \left(-\frac{1}{4}\right)\left(-\frac{1}{u^2}\right)(4x^3 + 12x)$

$= \frac{4x^3 + 12x}{4u^2} = \frac{4(x^3 + 3x)}{4u^2} = \frac{x^3 + 3x}{u^2} = \frac{x^3 + 3x}{(x^4 + 6x^2)^2}$

⑲ $\frac{y^5}{\sqrt{y^2-5}} = \frac{y^4 y^1}{\sqrt{y^2-5}} = \frac{y^4 y}{\sqrt{y^2-5}}, \int \frac{y^5}{\sqrt{y^2-5}} dy = \int \frac{y^4 y}{\sqrt{y^2-5}} dy$

$u = y^2 - 5, \frac{du}{dy} = 2y, du = 2y\,dy, \frac{du}{2} = y\,dy$

$y^2 = u + 5, y^4 = (y^2)^2 = (u+5)^2 = u^2 + 10u + 25$

$\int \frac{y^5}{\sqrt{y^2-5}} dy = \int \frac{y^4 y}{\sqrt{y^2-5}} dy = \int \frac{y^4}{\sqrt{y^2-5}} y\,dy = \int \frac{u^2 + 10u + 25}{\sqrt{u}} \frac{du}{2}$

$= \frac{1}{2}\int \frac{u^2 + 10u + 25}{u^{1/2}} du = \frac{1}{2}\int (u^{3/2} + 10u^{1/2} + 25u^{-1/2}) du$

$= \frac{1}{2}\int u^{3/2} du + 5\int u^{1/2} du + \frac{25}{2}\int u^{-1/2} du = \frac{1}{2}\left(\frac{u^{\frac{3}{2}+1}}{\frac{3}{2}+1}\right) + 5\left(\frac{u^{\frac{1}{2}+1}}{\frac{1}{2}+1}\right) + \frac{25}{2}\left(\frac{u^{-\frac{1}{2}+1}}{-\frac{1}{2}+1}\right) + c$

$= \frac{1}{2}\left(\frac{u^{5/2}}{5/2}\right) + 5\left(\frac{u^{3/2}}{3/2}\right) + \frac{25}{2}\left(\frac{u^{1/2}}{1/2}\right) + c = \frac{1}{2}\frac{2}{5}u^{5/2} + 5\frac{2}{3}u^{3/2} + \frac{25}{2}2u^{1/2} + c$

$= \frac{u^{5/2}}{5} + \frac{10u^{3/2}}{3} + 25u^{1/2} + c = \frac{(y^2-5)^{5/2}}{5} + \frac{10(y^2-5)^{3/2}}{3} + 25(y^2-5)^{1/2} + c$

Notes: $\frac{u^2}{u^{1/2}} = u^{2-1/2} = u^{3/2}, \frac{10u}{u^{1/2}} = 10u^{1-1/2} = 10u^{1/2}, \frac{25}{u^{1/2}} = 25u^{-1/2}$.

Check the answer: $f = \frac{u^{5/2}}{5} + \frac{10u^{3/2}}{3} + 25u^{1/2} + c, u = y^2 - 5$

$\frac{df}{dy} = \frac{df}{du}\frac{du}{dy} = \frac{d}{du}\left(\frac{u^{5/2}}{5} + \frac{10u^{3/2}}{3} + 25u^{1/2} + c\right)\frac{d}{dy}(y^2 - 5) = \left(\frac{u^{3/2}}{2} + 5u^{1/2} + \frac{25u^{-1/2}}{2}\right)(2y)$

$= (u^{3/2} + 10u^{1/2} + 25u^{-1/2})(y) = (u^2 + 10u + 25)(yu^{-1/2}) = \frac{(u^2 + 10u + 25)y}{u^{1/2}}$

$= \frac{[(y^2-5)^2 + 10(y-5) + 25]y}{u^{1/2}} = \frac{(y^4 - 10y + 25 + 10y - 50 + 25)y}{u^{1/2}}$

$= \frac{(y^4 + 0 + 0)y}{u^{1/2}} = \frac{y^4 y}{u^{1/2}} = \frac{y^5}{u^{1/2}} = \frac{y^5}{\sqrt{y^2-5}}$

Note: We factored out a $u^{-1/2}$. Note, for example, that $u^2 u^{-1/2} = u^{3/2}$.

㉑ $w^5\sqrt{w^3+3} = w^3 w^2 \sqrt{w^3+3}, \int w^5\sqrt{w^3+3}\,dw = \int w^3 w^2 \sqrt{w^3+3}\,dw$

$u = w^3+3, \dfrac{du}{dw} = 3w^2, du = 3w^2\,dw, \dfrac{du}{3} = w^2\,dw, w^3 = u-3$

$\int w^5\sqrt{w^3+3}\,dw = \int w^3 w^2 \sqrt{w^3+3}\,dw = \int w^3 \sqrt{w^3+3}\, w^2\,dw = \int (u-3)\sqrt{u}\,\dfrac{du}{3}$

$= \dfrac{1}{3}\int (u-3)\sqrt{u}\,du = \dfrac{1}{3}\int (u-3)u^{1/2}\,du = \dfrac{1}{3}\int u^{3/2}\,du - \dfrac{1}{3}\int 3u^{1/2}\,du$

$= \dfrac{1}{3}\int u^{3/2}\,du - \int u^{1/2}\,du = \dfrac{1}{3}\left(\dfrac{u^{\frac{3}{2}+1}}{\frac{3}{2}+1}\right) - \dfrac{u^{\frac{1}{2}+1}}{\frac{1}{2}+1} + c = \dfrac{1}{3}\left(\dfrac{u^{5/2}}{5/2}\right) - \dfrac{u^{3/2}}{3/2} + c$

$= \dfrac{1}{3}\dfrac{2}{5}u^{5/2} - \dfrac{2}{3}u^{3/2} + c = \dfrac{2}{15}u^{5/2} - \dfrac{2}{3}u^{3/2} + c = \dfrac{2}{15}(w^3+3)^{5/2} - \dfrac{2}{3}(w^3+3)^{3/2} + c$

Check the answer: $f = \dfrac{2}{15}u^{5/2} - \dfrac{2}{3}u^{3/2} + c = \dfrac{2}{15}(w^3+3)^{5/2} - \dfrac{2}{3}(w^3+3)^{3/2} + c$,

$u = w^3+3$

$\dfrac{df}{dw} = \dfrac{df}{du}\dfrac{du}{dw} = \dfrac{d}{du}\left(\dfrac{2}{15}u^{5/2} - \dfrac{2}{3}u^{3/2} + c\right)\dfrac{d}{dw}(w^3+3)$

$= \left(\dfrac{1}{3}u^{3/2} - u^{1/2}\right)(3w^2) = \left(\dfrac{u}{3}-1\right)(3w^2 u^{1/2}) = \left(\dfrac{u}{3}-\dfrac{3}{3}\right)(3w^2 u^{1/2})$

$= \left(\dfrac{u-3}{3}\right)(3w^2 u^{1/2}) = \left(\dfrac{w^3}{3}\right)(3w^2 u^{1/2}) = w^5 u^{1/2} = w^5\sqrt{w^3+3}$

Notes: We factored out a $u^{1/2}$. For example, $uu^{1/2} = u^1 u^{1/2} = u^{3/2}$. In the last line, we used $u = w^3+3$ to get $u-3 = w^3$, which allows us to replace $u-3$ with w^3.

Answers with Full Solutions

Chapter 3 Solutions

① $\int \tan(2\theta)\, d\theta = \int \dfrac{\sin(2\theta)}{\cos(2\theta)}\, d\theta$

$u = \cos(2\theta),\ \dfrac{du}{d\theta} = -2\sin(2\theta),\ du = -2\sin(2\theta)\, d\theta,\ -\dfrac{du}{2} = \sin(2\theta)\, d\theta$

$\int \dfrac{\sin(2\theta)}{\cos(2\theta)}\, d\theta = -\int \dfrac{du}{2u} = -\dfrac{\ln|u|}{2} + c = -\dfrac{\ln|\cos(2\theta)|}{2} + c = \dfrac{\ln|\sec(2\theta)|}{2} + c$

Note: We used the identity $-\ln w = \ln\left(\dfrac{1}{w}\right)$. That is, $-\dfrac{\ln|\cos(2\theta)|}{2} = \dfrac{1}{2}\ln\left|\dfrac{1}{\cos(2\theta)}\right| = \dfrac{\ln|\sec(2\theta)|}{2}$.

Alternate answers: $\ln\sqrt{\sec(2\theta)} + c$, $-\dfrac{\ln|\cos^2\theta - \sin^2\theta|}{2} + c$, etc. For the first alternate, we used the identity $c\ln w = \ln w^c$ to write $\dfrac{1}{2}\ln w = \ln(w^{1/2}) = \ln\sqrt{w}$.

Check the answer: $f = -\dfrac{\ln u}{2} + c = -\dfrac{\ln[\cos(2\theta)]}{2} + c,\ u = \cos y,\ y = 2\theta$

$\dfrac{df}{d\theta} = \dfrac{df}{du}\dfrac{du}{dy}\dfrac{dy}{d\theta} = \dfrac{d}{du}\left(-\dfrac{\ln u}{2} + c\right)\dfrac{d}{dy}\cos y \dfrac{dy}{d\theta}(2\theta) = \left(-\dfrac{1}{2u}\right)(-\sin y)(2)$

$= \dfrac{\sin y}{u} = \dfrac{\sin(2\theta)}{\cos(2\theta)} = \tan(2\theta)$

Note: The **chain rule** is used to check the answers for this chapter, much like we used the chain rule to check the answers to the examples.

② $u = 1 + \sin x,\ \dfrac{du}{dx} = \cos x,\ du = \cos x\, dx$

$\int \dfrac{\cos x}{1 + \sin x}\, dx = \int \dfrac{du}{u} = \ln|u| + c = \ln|1 + \sin x| + c$

Check the answer: $f = \ln u + c = \ln(1 + \sin x) + c,\ u = 1 + \sin x$

$\dfrac{df}{dx} = \dfrac{df}{du}\dfrac{du}{dx} = \dfrac{d}{du}(\ln u + c)\dfrac{d}{dx}(1 + \sin x) = \left(\dfrac{1}{u}\right)(\cos x) = \dfrac{\cos x}{u} = \dfrac{\cos x}{1 + \sin x}$

③ $u = \sin\varphi,\ \dfrac{du}{d\varphi} = \cos\varphi,\ du = \cos\varphi\, d\varphi$

At $\varphi = 0,\ u = \sin(0) = 0$. At $\varphi = \dfrac{\pi}{6},\ u = \sin\dfrac{\pi}{6} = \dfrac{1}{2}$.

$\displaystyle\int_{\varphi=0}^{\pi/6} \sin^2\varphi \cos\varphi\, d\varphi = \int_{u=0}^{1/2} u^2\, du = \left[\dfrac{u^3}{3}\right]_{u=0}^{1/2} = \dfrac{1}{3}\left(\dfrac{1}{2}\right)^3 - \dfrac{1}{3}(0)^3 = \dfrac{1}{3}\left(\dfrac{1}{8}\right) = \dfrac{1}{24} \approx 0.0417$

Techniques of Integration Calculus Practice Workbook

Check the antiderivative: $f = \frac{u^3}{3} = \frac{\sin^3 \varphi}{3}, u = \sin \varphi$

$\frac{df}{d\varphi} = \frac{df}{du}\frac{du}{d\varphi} = \frac{d}{du}\frac{u^3}{3}\frac{d}{d\varphi}\sin\varphi = u^2 \cos\varphi = \sin^2\varphi \cos\varphi$

④ $\cos^2\theta = \frac{1+\cos(2\theta)}{2}$, $\int_{\theta=0}^{\pi} \cos^2\theta\, d\theta = \int_{\theta=0}^{\pi} \frac{1+\cos(2\theta)}{2}\, d\theta$

$= \frac{1}{2}\int_{\theta=0}^{\pi} d\theta + \frac{1}{2}\int_{\theta=0}^{\pi} \cos(2\theta)\, d\theta$

$u = 2\theta, \frac{du}{d\theta} = 2, du = 2d\theta, \frac{du}{2} = d\theta$

At $\theta = 0, u = 2(0) = 0$. At $\theta = \pi, u = 2(\pi) = 2\pi$.

$\frac{1}{2}\int_{\theta=0}^{\pi} d\theta + \frac{1}{2}\int_{\theta=0}^{\pi} \cos(2\theta)\, d\theta = \frac{1}{2}[\theta]_{\theta=0}^{\pi} + \frac{1}{2}\int_{u=0}^{2\pi} \frac{\cos u}{2}\, du = \frac{\pi - 0}{2} + \frac{1}{4}\int_{u=0}^{2\pi} \cos u\, du$

$= \frac{\pi}{2} + \frac{1}{4}[\sin u]_{u=0}^{2\pi} = \frac{\pi}{2} + \frac{1}{4}(\sin 2\pi - \sin 0) = \frac{\pi}{2} + \frac{1}{4}(0 - 0) = \frac{\pi}{2} + 0 = \frac{\pi}{2} \approx 1.57$

Check the antiderivative: $f = \frac{\theta}{2} + \frac{\sin(2\theta)}{4}$. Apply the chain rule (Example 1) to the second term.

Note: Be sure to use $\frac{\theta}{2} + \frac{\sin u}{4}$, and **not** $\frac{\pi}{2} + \frac{\sin u}{4}$. You have to look back to $\frac{1}{2}[\theta]_{\theta=0}^{\pi}$ to see that the first term of the antiderivative was $\frac{\theta}{2}$ (before plugging in numbers).

$\frac{df}{d\theta} = \frac{d}{d\theta}\left[\frac{\theta}{2} + \frac{\sin(2\theta)}{4}\right] = \frac{1}{2} + \frac{\cos(2\theta)}{4}\frac{d}{d\theta}(2\theta) = \frac{1}{2} + \frac{\cos(2\theta)}{4}(2) = \frac{1}{2} + \frac{\cos(2\theta)}{2}$

$= \frac{1+\cos(2\theta)}{2} = \cos^2\theta$

⑤ $\int \cos^3 x\, dx = \int \cos^2 x \cos x\, dx = \int (1 - \sin^2 x) \cos x\, dx$

$= \int \cos x\, dx - \int \sin^2 x \cos x\, dx = \sin x - \int \sin^2 x \cos x\, dx$

$u = \sin x, \frac{du}{dx} = \cos x, du = \cos x\, dx$

$\sin x - \int \sin^2 x \cos x\, dx = \sin x - \int u^2\, du = \sin x - \frac{u^3}{3} + c = \sin x - \frac{\sin^3 x}{3} + c$

Alternate answer: $\frac{1}{12}\sin(3x) + \frac{3}{4}\sin x + c$

Answers with Full Solutions

Check the answer: $f = \sin x - \frac{\sin^3 x}{3} + c$. Apply the chain rule (Example 1) to the second term.

$$\frac{df}{dx} = \frac{d}{dx}\left(\sin x - \frac{\sin^3 x}{3} + c\right) = \cos x - \frac{3\sin^2 x}{3}\frac{d}{dx}\sin x = \cos x - \sin^2 x \cos x$$

$$= \cos x(1 - \sin^2 x) = \cos x \cos^2 x = \cos^3 x$$

6 $\sin(2y) = 2\sin y \cos y$, $\sin^2(2y) = 4\sin^2 y \cos^2 y$

$$\int \sin^2 y \cos^2 y \, dy = \int \frac{\sin^2(2y)}{4} dy = \frac{1}{4}\int \sin^2(2y) \, dy$$

$$\sin^2(2y) = \frac{1 - \cos(4y)}{2}$$

$$\frac{1}{4}\int \sin^2(2y) \, dy = \frac{1}{4}\int \frac{1 - \cos(4y)}{2} dy = \frac{1}{8}\int [1 - \cos(4y)] \, dy$$

$$= \frac{1}{8}\int dy - \frac{1}{8}\int \cos(4y) \, dy = \frac{y}{8} - \frac{1}{8}\int \cos(4y) \, dy$$

$$u = 4y, \frac{du}{dy} = 4, du = 4dy, \frac{du}{4} = dy$$

$$\frac{y}{8} - \frac{1}{8}\int \cos(4y) \, dy = \frac{y}{8} - \frac{1}{8}\int \frac{\cos(u)}{4} du = \frac{y}{8} - \frac{1}{32}\int \cos u \, du$$

$$= \frac{y}{8} - \frac{\sin u}{32} + c = \frac{y}{8} - \frac{\sin(4y)}{32} + c = \frac{y}{8} - \frac{\sin(2y)\cos(2y)}{16} + c$$

$$= \frac{y}{8} - \frac{\sin y \cos y (\cos^2 y - \sin^2 y)}{8} + c = \frac{y}{8} - \frac{\sin y \cos^3 y}{8} + \frac{\sin^3 y \cos y}{8} + c$$

Check the answer: $f = \frac{y}{8} - \frac{\sin y \cos^3 y}{8} + \frac{\sin^3 y \cos y}{8} + c$. Apply the **product rule** and the chain rule (Example 1) to the second and third terms.

$$\frac{d}{dy}\left(\frac{y}{8} - \frac{\sin y \cos^3 y}{8} + \frac{\sin^3 y \cos y}{8} + c\right)$$

$$= \frac{1}{8} - \frac{\cos^4 y}{8} - \frac{3\sin y \cos^2 y}{8}\frac{d}{dy}\cos y + \frac{3\sin^2 y \cos y}{8}\frac{d}{dy}\sin y - \frac{\sin^4 y}{8}$$

$$= \frac{1}{8} - \frac{\cos^4 y}{8} + \frac{3\sin^2 y \cos^2 y}{8} + \frac{3\sin^2 y \cos^2 y}{8} - \frac{\sin^4 y}{8}$$

$$= \frac{1}{8} + \frac{3\sin^2 y \cos^2 y}{4} - \frac{\cos^4 y}{8} - \frac{\sin^4 y}{8} = \frac{1}{8} + \frac{3\sin^2 y \cos^2 y}{4} - \frac{\cos^2 y \cos^2 y}{8} - \frac{\sin^2 y \sin^2 y}{8}$$

$$= \frac{1}{8} + \frac{3\sin^2 y \cos^2 y}{4} - \frac{\cos^2 y (1 - \sin^2 y)}{8} - \frac{\sin^2 y (1 - \cos^2 y)}{8}$$

$$= \frac{1}{8} + \frac{3\sin^2 y \cos^2 y}{4} - \frac{\cos^2 y}{8} + \frac{\sin^2 y \cos^2 y}{8} - \frac{\sin^2 y}{8} + \frac{\sin^2 y \cos^2 y}{8}$$

$$= \frac{1}{8} - \frac{\sin^2 y}{8} - \frac{\cos^2 y}{8} + \frac{3\sin^2 y \cos^2 y}{4} + \frac{\sin^2 y \cos^2 y}{8} + \frac{\sin^2 y \cos^2 y}{8}$$

$$= \frac{1}{8} - \frac{\sin^2 y + \cos^2 y}{8} + \frac{6\sin^2 y \cos^2 y}{8} + \frac{\sin^2 y \cos^2 y}{8} + \frac{\sin^2 y \cos^2 y}{8}$$

$$= \frac{1}{8} - \frac{1}{8} + \frac{8\sin^2 y \cos^2 y}{8} = 0 + \sin^2 y \cos^2 y = \sin^2 y \cos^2 y$$

7 $u = \sin t, \dfrac{du}{dt} = \cos t, du = \cos t\, dt$

At $t = \dfrac{\pi}{6}, u = \sin\dfrac{\pi}{6} = \dfrac{1}{2}$. At $t = \pi, u = \sin\pi = 0$.

$$\int_{t=\pi/6}^{\pi} \sin^3 t \cos t\, dt = \int_{u=1/2}^{0} u^3\, du = \left[\frac{u^4}{4}\right]_{u=1/2}^{0} = \frac{0^4}{4} - \frac{1}{4}\left(\frac{1}{2}\right)^4 = 0 - \frac{1}{4}\left(\frac{1}{16}\right)$$

$$= -\frac{1}{64} = -0.015625$$

Check the antiderivative: $f = \dfrac{u^4}{4}, u = \sin t$

$$\frac{df}{dt} = \frac{df}{du}\frac{du}{dt} = \frac{d}{du}\frac{u^4}{4}\frac{d}{dt}\sin t = u^3 \cos t = \sin^3 t \cos t$$

8 $\displaystyle\int_{\theta=0}^{\pi/3} \sin\theta \sec^2\theta\, d\theta = \int_{\theta=0}^{\pi/3} \frac{\sin\theta}{\cos^2\theta}\, d\theta$

$u = \cos\theta, \dfrac{du}{d\theta} = -\sin\theta, du = -\sin\theta\, d\theta$

At $\theta = 0, u = \cos 0 = 1$. At $\theta = \dfrac{\pi}{3}, u = \cos\dfrac{\pi}{3} = \dfrac{1}{2}$.

$$\int_{\theta=0}^{\pi/3} \frac{\sin\theta}{\cos^2\theta}\, d\theta = -\int_{u=1}^{1/2} \frac{du}{u^2} = -\int_{u=1}^{1/2} u^{-2}\, du = -\left[\frac{1}{-u}\right]_{u=1}^{1/2} = \left[\frac{1}{u}\right]_{u=1}^{1/2}$$

$$= \frac{1}{1/2} - \frac{1}{1} = 2 - 1 = 1 \quad \text{Simpler solution: } \int \sin\theta \sec^2\theta\, d\theta = \int \sec\theta \tan\theta\, d\theta = \sec\theta$$

Check the antiderivative: $f = \dfrac{1}{u} = \dfrac{1}{\cos\theta}, u = \cos\theta$

$$\frac{df}{d\theta} = \frac{df}{du}\frac{du}{d\theta} = \frac{d}{du}\frac{1}{u}\frac{d}{d\theta}\cos\theta = \left(-\frac{1}{u^2}\right)(-\sin\theta) = \frac{\sin\theta}{\cos^2\theta} = \sin\theta \sec^2\theta$$

Answers with Full Solutions

❾ $1 + \tan^2 \varphi = \sec^2 \varphi$, $\tan^2 \varphi = \sec^2 \varphi - 1$

$$\int \tan^2 \varphi \, d\varphi = \int \sec^2 \varphi \, d\varphi - \int d\varphi = \tan \varphi - \varphi + c = -\varphi + \tan \varphi + c$$

Check the answer: $f = \tan \varphi - \varphi + c = -\varphi + \tan \varphi + c$

$$\frac{d}{d\varphi}(-\varphi + \tan \varphi + c) = -1 + \sec^2 \varphi = \sec^2 \varphi - 1 = \tan^2 \varphi$$

❿ $\int \frac{\sec x \tan x}{\sqrt{\cos x}} dx = \int \sec x \tan x \sqrt{\frac{1}{\cos x}} dx = \int \sec x \tan x \sqrt{\sec x} \, dx$

$u = \sec x, \dfrac{du}{dx} = \sec x \tan x, du = \sec x \tan x \, dx$

$$\int \sec x \tan x \sqrt{\sec x} \, dx = \int \sqrt{u} \, du = \int u^{1/2} du = \frac{u^{\frac{1}{2}+1}}{\frac{1}{2}+1} + c = \frac{u^{3/2}}{3/2} + c$$

$$= \frac{2}{3} u^{3/2} + c = \frac{2}{3} \sec^{3/2} x + c = \frac{2}{3} \sec x \sqrt{\sec x} + c$$

Note: We used the rule from algebra that $u^{3/2} = u^1 u^{1/2} = u\sqrt{u}$.

Check the answer: $f = \frac{2}{3} u^{3/2} + c = \frac{2}{3} \sec^{3/2} x + c$, $u = \sec x$

$$\frac{df}{dx} = \frac{df}{du}\frac{du}{dx} = \frac{d}{du}\left(\frac{2}{3} u^{3/2} + c\right) \frac{d}{dx}\sec x = u^{1/2} \sec x \tan x = \sec x \tan x \sqrt{\sec x}$$

⓫ $u = \sec \theta, \dfrac{du}{d\theta} = \sec \theta \tan \theta, du = \sec \theta \tan \theta \, d\theta$

At $\theta = 0$, $u = \sec 0 = 1$. At $\theta = \dfrac{\pi}{3}$, $u = \sec \dfrac{\pi}{3} = 2$.

$$\int_{\theta=0}^{\pi/3} \sec^4 \theta \tan \theta \, d\theta = \int_{\theta=0}^{\pi/3} \sec^3 \theta \sec \theta \tan \theta \, d\theta = \int_{u=1}^{2} u^3 \, du = \left[\frac{u^4}{4}\right]_{u=1}^{2}$$

$$= \frac{2^4}{4} - \frac{1^4}{4} = \frac{16}{4} - \frac{1}{4} = \frac{15}{4} = 3.75$$

Check the antiderivative: $f = \dfrac{u^4}{4} = \dfrac{\sec^4 \theta}{4}$, $u = \sec \theta$

$$\frac{df}{d\theta} = \frac{df}{du}\frac{du}{d\theta} = \frac{d}{du}\frac{u^4}{4}\frac{d}{d\theta}\sec \theta = u^3 \sec \theta \tan \theta = \sec^3 \theta \sec \theta \tan \theta = \sec^4 \theta \tan \theta$$

⓬ $u = \tan w, \dfrac{du}{dw} = \sec^2 w, du = \sec^2 w \, dw$

At $w = 0$, $u = \tan 0 = 0$. At $\theta = \frac{\pi}{4}$, $u = \tan\frac{\pi}{4} = 1$.

$$\int_{w=0}^{\pi/4} \tan^3 w \sec^2 w \, dw = \int_{u=0}^{1} u^3 \, du = \left[\frac{u^4}{4}\right]_{u=0}^{1} = \frac{1^4}{4} - \frac{0^4}{4} = \frac{1}{4} - 0 = \frac{1}{4} = 0.25$$

Check the antiderivative: $f = \frac{u^4}{4} = \frac{\tan^4 w}{4}$, $u = \tan w$

$$\frac{df}{dw} = \frac{df}{du}\frac{du}{dw} = \frac{d}{du}\frac{u^4}{4}\frac{d}{dw}\tan w = u^3 \sec^2 w = \tan^3 w \sec^2 w$$

⑬ $x = 4\sin\theta$, $\frac{dx}{d\theta} = 4\cos\theta$, $dx = 4\cos\theta \, d\theta$ (see Example 9)

$$\int \sqrt{16 - x^2} \, dx = \int \sqrt{16 - (4\sin\theta)^2} \, 4\cos\theta \, d\theta = 4\int \cos\theta \sqrt{16 - 16\sin^2\theta} \, d\theta$$

$$= 4\int \cos\theta \sqrt{16(1-\sin^2\theta)} \, d\theta = 4\int \cos\theta \sqrt{16}\sqrt{1-\sin^2\theta} \, d\theta$$

$$= 4\int 4\cos\theta \sqrt{\cos^2\theta} \, d\theta = 16\int \cos^2\theta \, d\theta = 16\int \frac{1+\cos(2\theta)}{2} \, d\theta$$

$$= 8\int [1+\cos(2\theta)] \, d\theta = 8\int d\theta + 8\int \cos(2\theta) \, d\theta = 8\theta + 8\int \cos(2\theta) \, d\theta$$

$u = 2\theta$, $\frac{du}{d\theta} = 2$, $du = 2d\theta$, $\frac{du}{2} = d\theta$

$$8\theta + 8\int \cos(2\theta) \, d\theta = 8\theta + 8\int \frac{\cos u}{2} \, du = 8\theta + 4\int \cos u \, du$$

$$= 8\theta + 4\sin u + c = 8\theta + 4\sin(2\theta) + c = 8\theta + 8\sin\theta\cos\theta + c$$

$$= 8\sin^{-1}\left(\frac{x}{4}\right) + 8\sin\left[\sin^{-1}\left(\frac{x}{4}\right)\right]\cos\left[\sin^{-1}\left(\frac{x}{4}\right)\right] + c$$

$$= 8\sin^{-1}\left(\frac{x}{4}\right) + 8\left(\frac{x}{4}\right)\sqrt{1-\left(\frac{x}{4}\right)^2} + c = 8\sin^{-1}\left(\frac{x}{4}\right) + 2x\sqrt{1-\frac{x^2}{16}} + c$$

$$= 8\sin^{-1}\left(\frac{x}{4}\right) + 2x\sqrt{\frac{16-x^2}{16}} + c = 8\sin^{-1}\left(\frac{x}{4}\right) + \frac{2x}{\sqrt{16}}\sqrt{16-x^2} + c$$

$$= 8\sin^{-1}\left(\frac{x}{4}\right) + \frac{2x}{4}\sqrt{16-x^2} + c = 8\sin^{-1}\left(\frac{x}{4}\right) + \frac{x}{2}\sqrt{16-x^2} + c$$

Note: We used the trig identities $\sin(\sin^{-1} w) = w$ and $\cos(\sin^{-1} w) = \sqrt{1-w^2}$.[20]

---

[20] We want $\cos(\sin^{-1} w)$. The angle of the cosine function equals $\sin^{-1} w$, so the sine of that

Answers with Full Solutions

Check the answer: $f = 8\sin^{-1}\left(\frac{x}{4}\right) + \frac{x}{2}\sqrt{16-x^2} + c$ Apply the chain rule (Example 1) and recall from first-semester calculus that $\frac{d}{dx}\sin^{-1}g = \frac{1}{\sqrt{1-g^2}}\frac{dg}{dx}$ if $-\frac{\pi}{2} < \sin^{-1}x < \frac{\pi}{2}$.

Also, apply the **product rule**.

$$\frac{df}{dx} = \frac{d}{dx}\left[8\sin^{-1}\left(\frac{x}{4}\right) + \frac{x}{2}\sqrt{16-x^2} + c\right]$$

$$= \frac{8}{\sqrt{1-\left(\frac{x}{4}\right)^2}}\frac{d}{dx}\left(\frac{x}{4}\right) + \frac{1}{2}\sqrt{16-x^2} + \frac{x}{2}\frac{1}{2\sqrt{16-x^2}}\frac{d}{dx}(16-x^2)$$

$$= \frac{8}{\sqrt{1-\frac{x^2}{16}}}\left(\frac{1}{4}\right) + \frac{\sqrt{16-x^2}}{2}\frac{\sqrt{16-x^2}}{\sqrt{16-x^2}} + \frac{x}{2}\frac{1}{2\sqrt{16-x^2}}(-2x)$$

$$= \frac{2}{\sqrt{1-\frac{x^2}{16}}} + \frac{16-x^2}{2\sqrt{16-x^2}} - \frac{x^2}{2\sqrt{16-x^2}} = \frac{4(2)}{4\sqrt{1-\frac{x^2}{16}}} + \frac{16-x^2-x^2}{2\sqrt{16-x^2}}$$

$$= \frac{8}{\sqrt{16-x^2}} + \frac{16-2x^2}{2\sqrt{16-x^2}} = \frac{8}{\sqrt{16-x^2}} + \frac{8-x^2}{\sqrt{16-x^2}} = \frac{8+8-x^2}{\sqrt{16-x^2}} = \frac{16-x^2}{\sqrt{16-x^2}} = \sqrt{16-x^2}$$

⑭ $t = \sqrt{3}\tan\theta, \frac{dt}{d\theta} = \sqrt{3}\sec^2\theta, dt = \sqrt{3}\sec^2\theta\, d\theta$ (see Example 10)

$$\int \frac{dt}{t^2+3} = \int \frac{\sqrt{3}\sec^2\theta}{\left(\sqrt{3}\tan\theta\right)^2+3}d\theta = \sqrt{3}\int\frac{\sec^2\theta}{3\tan^2\theta+3}d\theta = \sqrt{3}\int\frac{\sec^2\theta}{3(\tan^2\theta+1)}d\theta$$

$$= \sqrt{3}\int\frac{\sec^2\theta}{3\sec^2\theta}d\theta = \frac{\sqrt{3}}{3}\int d\theta = \frac{\sqrt{3}}{3}\theta + c = \frac{\sqrt{3}}{3}\tan^{-1}\left(\frac{t}{\sqrt{3}}\right) + c = \frac{\sqrt{3}}{3}\tan^{-1}\left(\frac{t\sqrt{3}}{3}\right) + c$$

Alternate answer: $\frac{1}{\sqrt{3}}\tan^{-1}\left(\frac{t}{\sqrt{3}}\right) + c$

Note: The purpose of this exercise was to practice using trig substitutions. If you're fluent with your derivatives of inverse trig functions, this integral is trivial, but it would be good practice to use the above substitution instead.

Check the answer: $f = \frac{\sqrt{3}}{3}\tan^{-1}\left(\frac{t}{\sqrt{3}}\right) + c$ Recall from first-semester calculus that

angle equals w. Draw a right triangle where the opposite is w and the hypotenuse is 1 (so the sine is $\frac{w}{1} = w$). The adjacent will be $\sqrt{1-w^2}$. The cosine is adjacent over hypotenuse: $\frac{\sqrt{1-w^2}}{1} = \sqrt{1-w^2}$. That's one way to find that $\cos(\sin^{-1}w) = \sqrt{1-w^2}$.

162

$\frac{d}{dt}\tan^{-1} u = \frac{1}{1+u^2}\frac{du}{dt}$ if $-\frac{\pi}{2} < \tan^{-1} t < \frac{\pi}{2}$.

$\frac{df}{dt} = \frac{d}{dt}\left[\frac{\sqrt{3}}{3}\tan^{-1}\left(\frac{t}{\sqrt{3}}\right) + c\right] = \frac{\sqrt{3}}{3}\frac{1}{1+\left(\frac{t}{\sqrt{3}}\right)^2}\frac{d}{dt}\left(\frac{t}{\sqrt{3}}\right) = \frac{\sqrt{3}}{3}\frac{1}{1+\frac{t^2}{3}}\frac{1}{\sqrt{3}}$

$= \frac{1}{3}\frac{1}{1+\frac{t^2}{3}} = \frac{1}{3\left(1+\frac{t^2}{3}\right)} = \frac{1}{3+t^2}$

⑮ $y = \sec\theta, \frac{dy}{d\theta} = \sec\theta\tan\theta, dy = \sec\theta\tan\theta\, d\theta$ (see Example 11)

At $y = \frac{2}{\sqrt{3}}, \theta = \sec^{-1}\left(\frac{2}{\sqrt{3}}\right) = \frac{\pi}{6}$. At $y = 2, \theta = \sec^{-1} 2 = \frac{\pi}{3}$.

$\int_{y=2/\sqrt{3}}^{2} \frac{dy}{y^2 - 1} = \int_{\theta=\pi/6}^{\pi/3} \frac{\sec\theta\tan\theta}{\sec^2\theta - 1} d\theta = \int_{\theta=\pi/6}^{\pi/3} \frac{\sec\theta\tan\theta}{\tan^2\theta} d\theta = \int_{\theta=\pi/6}^{\pi/3} \frac{\sec\theta}{\tan\theta} d\theta$

$= \int_{\theta=\pi/6}^{\pi/3} \csc\theta\, d\theta = [\ln|\csc\theta - \cot\theta|]_{\theta=\pi/6}^{\pi/3} = \ln\left|\csc\frac{\pi}{3} - \cot\frac{\pi}{3}\right| - \ln\left|\csc\frac{\pi}{6} - \cot\frac{\pi}{6}\right|$

$= \ln\left|\frac{2}{\sqrt{3}} - \frac{1}{\sqrt{3}}\right| - \ln|2 - \sqrt{3}| = \ln\left|\frac{1}{\sqrt{3}}\right| - \ln|2 - \sqrt{3}| = \ln\left|\frac{1}{\sqrt{3}} \div 2 - \sqrt{3}\right|$

$= \ln\left|\frac{1}{\sqrt{3}(2-\sqrt{3})}\right| = \ln\left|\frac{1}{2\sqrt{3}-3}\right| \approx 0.768$ (see the paragraph that follows)

Check the antiderivative: $f = \ln[\csc(\sec^{-1} y) - \cot(\sec^{-1} y)] = \ln\left(\frac{y}{\sqrt{y^2-1}} - \frac{1}{\sqrt{y^2-1}}\right)$

$= \ln\left(\frac{y-1}{\sqrt{y^2-1}}\right) = \ln\left(\frac{\sqrt{(y-1)^2}}{\sqrt{y^2-1}}\right) = \ln\sqrt{\frac{(y-1)^2}{y^2-1}} = \frac{1}{2}\ln\frac{(y-1)^2}{y^2-1} = \frac{1}{2}\ln\frac{(y-1)^2}{(y+1)(y-1)} = \frac{1}{2}\ln\left(\frac{y-1}{y+1}\right)$.

This answer, $\frac{1}{2}\ln\left(\frac{y-1}{y+1}\right)$, **is the common form** of $\int \frac{dy}{y^2-1}$ (when $y^2 > 1$). If you're fluent in your inverse hyperbolic trig functions, it equates to $-\coth^{-1} y$. If you use the common form to evaluate the definite integral, it's tricky to express the answer in the same algebraic form $\ln\left|\frac{1}{2\sqrt{3}-3}\right|$ (but it's also a good algebra challenge; you can find the solution on the next page, in case you wish to try it), though reasonable to see that it approximately equals 0.768. Above, we used the trig identities $\csc(\sec^{-1} y) = \frac{y}{\sqrt{y^2-1}}$ and $\cot(\sec^{-1} y) = \frac{1}{\sqrt{y^2-1}}$. See Footnotes 7 and 9.

Answers with Full Solutions

$$\left[\frac{1}{2}\ln\left(\frac{y-1}{y+1}\right)\right]^2_{y=2/\sqrt{3}} = \frac{1}{2}\ln\left(\frac{2-1}{2+1}\right) - \frac{1}{2}\ln\left(\frac{\frac{2}{\sqrt{3}}-1}{\frac{2}{\sqrt{3}}+1}\right) = \frac{1}{2}\ln\left(\frac{1}{3}\right) - \frac{1}{2}\ln\left(\frac{2-\sqrt{3}}{2+\sqrt{3}}\right)$$

$$= \frac{1}{2}\ln\left(\frac{1}{3}\right) - \frac{1}{2}\ln\left[\frac{2-\sqrt{3}}{2+\sqrt{3}}\left(\frac{2-\sqrt{3}}{2-\sqrt{3}}\right)\right] = \frac{1}{2}\ln\left(\frac{1}{3}\right) - \frac{1}{2}\ln\left[\frac{4-2\sqrt{3}-2\sqrt{3}+3}{4-2\sqrt{3}+2\sqrt{3}-3}\right]$$

$$= \frac{1}{2}\ln\left(\frac{1}{3}\right) - \frac{1}{2}\ln\left(\frac{7-4\sqrt{3}}{1}\right) = \frac{1}{2}\ln\left(\frac{1}{3}\right) - \frac{1}{2}\ln(7-4\sqrt{3}) = \ln\left(\frac{1}{\sqrt{3}}\right) - \ln\sqrt{7-4\sqrt{3}}$$

$$= \ln\left(\frac{1}{\sqrt{3}}\right) - \ln(2-\sqrt{3}) = \ln\left[\frac{1}{\sqrt{3}} \div (2-\sqrt{3})\right] = \ln\left[\frac{1}{\sqrt{3}(2-\sqrt{3})}\right] = \ln\left(\frac{1}{2\sqrt{3}-3}\right)$$

We used the identities $\ln p - \ln q = \ln \frac{p}{q}$ and $\frac{1}{2}\ln r = \ln r^{1/2} = \ln \sqrt{r}$. **The trickiest part of this algebra** is that $\sqrt{7-4\sqrt{3}} = 2-\sqrt{3}$. The easy way to convince yourself that this is true is to multiply $2-\sqrt{3}$ by itself.

To check the antiderivative, apply the chain rule (Example 1) and **quotient rule**.

$$\frac{df}{dy} = \frac{d}{dy}\left[\frac{1}{2}\ln\left(\frac{y-1}{y+1}\right)\right] = \frac{1}{2}\frac{1}{(y-1)/(y+1)}\frac{d}{dy}\frac{y-1}{y+1}$$

$$= \left(\frac{y+1}{y-1}\right)\frac{(y+1)\frac{d}{dy}(y-1)-(y-1)\frac{d}{dy}(y+1)}{(y+1)^2} = \frac{1}{2}\left(\frac{y+1}{y-1}\right)\frac{(y+1)-(y-1)}{(y+1)^2}$$

$$= \frac{1}{2}\left(\frac{y+1}{y-1}\right)\frac{y+1-y-(-1)}{(y+1)^2} = \frac{1}{2}\left(\frac{y+1}{y-1}\right)\frac{y+1-y+1}{(y+1)^2} = \frac{1}{2}\left(\frac{y+1}{y-1}\right)\frac{2}{(y+1)^2}$$

$$= \frac{1}{(y-1)(y+1)} = \frac{1}{y^2-1}$$

⑯ $x = 5\tan\theta, \frac{dx}{d\theta} = 5\sec^2\theta, dx = 5\sec^2\theta\, d\theta$ (see Example 10)

At $x=0, \theta = \tan^{-1}\left(\frac{0}{5}\right) = 0$. At $x=5, \theta = \tan^{-1}\left(\frac{5}{5}\right) = \frac{\pi}{4}$.

$$\int_{x=0}^{5}\frac{x^2}{x^2+25}dx = \int_{\theta=0}^{\pi/4}\frac{(5\tan\theta)^2}{(5\tan\theta)^2+25}5\sec^2\theta\, d\theta = \int_{\theta=0}^{\pi/4}\frac{25\tan^2\theta\, 5\sec^2\theta}{25\tan^2\theta+25}d\theta$$

$$= \int_{\theta=0}^{\pi/4}\frac{125\tan^2\theta\sec^2\theta}{25(\tan^2\theta+1)}d\theta = 5\int_{\theta=0}^{\pi/4}\frac{\tan^2\theta\sec^2\theta}{\sec^2\theta}d\theta = 5\int_{\theta=0}^{\pi/4}\tan^2\theta\, d\theta$$

$$= 5 \int_{\theta=0}^{\pi/4} (\sec^2 \theta - 1) \, d\theta = 5 \int_{\theta=0}^{\pi/4} \sec^2 \theta \, d\theta - 5 \int_{\theta=0}^{\pi/4} d\theta = 5[\tan \theta]_{\theta=0}^{\pi/4} - 5[\theta]_{\theta=0}^{\pi/4}$$

$$= 5 \tan\left(\frac{\pi}{4}\right) - 5 \tan 0 - 5\left(\frac{\pi}{4}\right) + 5(0) = 5(1) - 5(0) - \frac{5\pi}{4} + 0$$

$$= 5 - \frac{5\pi}{4} = \frac{20}{4} - \frac{5\pi}{4} = \frac{20 - 5\pi}{4} = 5\left(\frac{4-\pi}{4}\right) = 5\left(1 - \frac{\pi}{4}\right) \approx 1.073$$

Check the antiderivative: $f = 5 \tan \theta - 5\theta = 5 \tan\left[\tan^{-1}\left(\frac{x}{5}\right)\right] - 5 \tan^{-1}\left(\frac{x}{5}\right)$

$= \frac{5x}{5} - 5 \tan^{-1}\left(\frac{x}{5}\right) = x - 5 \tan^{-1}\left(\frac{x}{5}\right)$. Recall from first-semester calculus that

$\frac{d}{dx} \tan^{-1} u = \frac{1}{1+u^2} \frac{du}{dx}$ if $-\frac{\pi}{2} < \tan^{-1} x < \frac{\pi}{2}$.

$\frac{df}{dx} = \frac{d}{dx}\left[x - 5 \tan^{-1}\left(\frac{x}{5}\right)\right] = 1 - 5 \frac{1}{1+\left(\frac{x}{5}\right)^2} \frac{d}{dx}\left(\frac{x}{5}\right) = 1 - \frac{5}{1+\frac{x^2}{25}}\left(\frac{1}{5}\right) = 1 - \frac{1}{\frac{25}{25} + \frac{x^2}{25}}$

$= 1 - \frac{1}{\frac{25+x^2}{25}} = 1 - \frac{25}{25+x^2} = \frac{25+x^2}{25+x^2} - \frac{25}{25+x^2} = \frac{x^2}{25+x^2}$

17 $u = \sin \theta, \frac{du}{d\theta} = \cos \theta, du = \cos \theta \, d\theta$

$$\int \frac{\cos \theta}{\sqrt{1+\sin^2 \theta}} d\theta = \int \frac{du}{\sqrt{1+u^2}}$$

$u = \tan w, \frac{du}{dw} = \sec^2 w, du = \sec^2 w \, dw$ (see Example 10)

$$\int \frac{du}{\sqrt{1+u^2}} = \int \frac{\sec^2 w \, dw}{\sqrt{1+\tan^2 w}} = \int \frac{\sec^2 w \, dw}{\sqrt{\sec^2 w}} = \int \frac{\sec^2 w \, dw}{\sec w} = \int \sec w \, dw$$

$= \ln|\sec w + \tan w| + c = \ln|\sec(\tan^{-1} u) + \tan(\tan^{-1} u)| + c$

$= \ln\left|\sqrt{1+u^2} + u\right| + c = \ln\left|u + \sqrt{1+u^2}\right| + c = \ln\left|\sin \theta + \sqrt{1+\sin^2 \theta}\right| + c$

Note: We used the identity $\sec(\tan^{-1} u) = \sqrt{1+u^2}$. See Footnotes 7 and 9.

Check the answer: $f = \ln\left(\sin \theta + \sqrt{1+\sin^2 \theta}\right) + c$. Apply the chain rule (Example 1.)

$\frac{df}{d\theta} = \frac{d}{d\theta}\left[\ln\left(\sin \theta + \sqrt{1+\sin^2 \theta}\right) + c\right]$

$= \frac{1}{\sin \theta + \sqrt{1+\sin^2 \theta}}\left[\cos \theta + \frac{1}{2\sqrt{1+\sin^2 \theta}} \frac{d}{d\theta}(1+\sin^2 \theta)\right]$

Answers with Full Solutions

$$= \frac{1}{\sin\theta + \sqrt{1+\sin^2\theta}}\left(\cos\theta + \frac{1}{2\sqrt{1+\sin^2\theta}} 2\sin\theta \frac{d}{d\theta}\sin\theta\right)$$

$$= \frac{1}{\sin\theta + \sqrt{1+\sin^2\theta}}\left(\cos\theta + \frac{\sin\theta\cos\theta}{\sqrt{1+\sin^2\theta}}\right) = \frac{\cos\theta}{\sin\theta + \sqrt{1+\sin^2\theta}}\left(1 + \frac{\sin\theta}{\sqrt{1+\sin^2\theta}}\right)$$

$$= \frac{\cos\theta}{\sin\theta + \sqrt{1+\sin^2\theta}}\left(\frac{\sqrt{1+\sin^2\theta}}{\sqrt{1+\sin^2\theta}} + \frac{\sin\theta}{\sqrt{1+\sin^2\theta}}\right)$$

$$= \frac{\cos\theta}{\sin\theta + \sqrt{1+\sin^2\theta}}\left(\frac{\sin\theta + \sqrt{1+\sin^2\theta}}{\sqrt{1+\sin^2\theta}}\right) = \frac{\cos\theta}{\sqrt{1+\sin^2\theta}}$$

⑱ Note: The 'trick' is to think of $t^4 = (t^2)^2$. Working with the square of t^2, the substitution $u = t^2$ leads to $du = 2t\,dt$, and the numerator has this t.

$$u = t^2, \frac{du}{dt} = 2t, du = 2t\,dt, \frac{du}{2} = t\,dt \quad \text{Note: } u^2 = (t^2)^2 = t^4$$

$$\int \frac{t}{(1-t^4)^{3/2}} dt = \int \frac{du}{2(1-u^2)^{3/2}} = \frac{1}{2}\int \frac{du}{(1-u^2)^{3/2}}$$

$$u = \sin\theta, \frac{du}{d\theta} = \cos\theta, du = \cos\theta\,d\theta \quad \text{(see Example 9)}$$

$$\frac{1}{2}\int \frac{du}{(1-u^2)^{3/2}} = \frac{1}{2}\int \frac{\cos\theta}{(1-\sin^2\theta)^{3/2}} d\theta = \frac{1}{2}\int \frac{\cos\theta}{(\cos^2\theta)^{3/2}} d\theta = \frac{1}{2}\int \frac{\cos\theta}{\cos^3\theta} d\theta$$

$$= \frac{1}{2}\int \frac{d\theta}{\cos^2\theta} = \frac{1}{2}\int \sec^2\theta\,d\theta = \frac{1}{2}\tan\theta + c = \frac{1}{2}\tan(\sin^{-1}u) + c$$

$$= \frac{1}{2}\frac{u}{\sqrt{1-u^2}} + c = \frac{t^2}{2\sqrt{1-t^4}} + c$$

Note: We used the identity $\tan(\sin^{-1}u) = \frac{u}{\sqrt{1-u^2}}$. See Footnotes 7 and 9.

Check the answer: $f = \frac{t^2}{2\sqrt{1-t^4}} + c$. Apply the **quotient rule** and chain rule (Example 1).

$$\frac{df}{dt} = \frac{d}{dt}\left(\frac{t^2}{2\sqrt{1-t^4}} + c\right) = \frac{1}{2}\frac{\sqrt{1-t^4}\frac{d}{dt}t^2 - t^2\frac{d}{dt}\sqrt{1-t^4}}{\left(\sqrt{1-t^4}\right)^2}$$

$$= \frac{1}{2}\frac{\sqrt{1-t^4}(2t) - t^2\frac{1}{2\sqrt{1-t^4}}\frac{d}{dt}(1-t^4)}{(1-t^4)} = \frac{2t\sqrt{1-t^4} - \frac{t^2}{2\sqrt{1-t^4}}(-4t^3)}{2(1-t^4)}$$

$$= \frac{2t\sqrt{1-t^4} + \frac{2t^5}{\sqrt{1-t^4}}}{2(1-t^4)} = \frac{2t\sqrt{1-t^4}\frac{\sqrt{1-t^4}}{\sqrt{1-t^4}} + \frac{2t^5}{\sqrt{1-t^4}}}{2(1-t^4)} = \frac{\frac{2t(1-t^4)}{\sqrt{1-t^4}} + \frac{2t^5}{\sqrt{1-t^4}}}{2(1-t^4)}$$

$$= \frac{\frac{2t-2t^5}{\sqrt{1-t^4}} + \frac{2t^5}{\sqrt{1-t^4}}}{2(1-t^4)} = \frac{\frac{2t}{\sqrt{1-t^4}}}{2(1-t^4)} = \frac{2t}{\sqrt{1-t^4}} \div [2(1-t^4)]$$

$$= \frac{2t}{\sqrt{1-t^4}} \times \frac{1}{2(1-t^4)} = \frac{t}{(1-t^4)^{3/2}}$$

⑲ $u = 1 + \sin^2 x, \frac{du}{dx} = 2\sin x \frac{d}{dx}\sin x = 2\sin x \cos x = \sin(2x), du = \sin(2x)\,dx$

Note: We used the chain rule (Example 1) with $u = 1 + z^2$ and $z = \sin x$ to get

$$\frac{du}{dx} = \frac{du}{dz}\frac{dz}{dx} = \frac{d}{dz}(1+z^2)\frac{d}{dx}\sin x = 2z\cos x = 2\sin x \cos x = \sin(2x)$$

such that $du = \sin(2x)\,dx$.

At $x = 0, u = 1 + \sin^2 0 = 1 + 0 = 1$. At $x = \frac{\pi}{2}, u = 1 + \sin^2\left(\frac{\pi}{2}\right) = 1 + 1^2 = 2$.

$$\int_{x=0}^{\pi/2} \frac{\sin(2x)}{1+\sin^2 x}dx = \int_{u=1}^{2}\frac{du}{u} = [\ln|u|]_{u=1}^{2} = \ln 2 - \ln 1 = \ln 2 - 0 = \ln 2 \approx 0.693$$

Check the antiderivative: $f = \ln u = \ln(1+z^2) = \ln(1+\sin^2 x)$. Apply the chain rule (Example 1) with $u = 1 + z^2 = 1 + \sin^2 x$ and $z = \sin x$.

$$\frac{df}{dx} = \frac{df}{du}\frac{du}{dz}\frac{dz}{dx} = \frac{d}{du}\ln u \frac{d}{dz}(1+z^2)\frac{d}{dx}\sin x = \frac{1}{u}(2z)\cos x = \frac{2z\cos x}{u}$$

$$= \frac{2z\cos x}{1+z^2} = \frac{2\sin x \cos x}{1+\sin^2 x} = \frac{\sin(2x)}{1+\sin^2 x}$$

⑳ The 'trick' is to recall from first-semester calculus that $\frac{d}{dy}\tan^{-1} y = \frac{1}{1+y^2}$ if $-\frac{\pi}{2} < \tan^{-1} y < \frac{\pi}{2}$.

$u = \tan^{-1} y, \frac{du}{dy} = \frac{1}{1+y^2}, du = \frac{dy}{1+y^2}$

At $y = 0, u = \tan^{-1} 0 = 0$. At $y = 1, u = \tan^{-1} 1 = \frac{\pi}{4}$.

$$\int_{y=0}^{1}\frac{\tan^{-1} y}{1+y^2}dy = \int_{y=0}^{1}\tan^{-1} y\frac{dy}{1+y^2} = \int_{u=0}^{\pi/4} u\,du = \left[\frac{u^2}{2}\right]_{u=0}^{\pi/4} = \frac{1}{2}\left(\frac{\pi}{4}\right)^2 - \frac{0^2}{2}$$

$$= \frac{1}{2}\left(\frac{\pi^2}{16}\right) - 0 = \frac{\pi^2}{32} \approx 0.308$$

Note: Occasionally, the hardest-looking problems turn out to be among the simplest. Imagine all the students who glanced at this problem and decided to skip it.

Check the antiderivative: $f = \frac{u^2}{2} = \frac{(\tan^{-1} y)^2}{2}, u = \tan^{-1} y$

$\frac{df}{dy} = \frac{df}{du}\frac{du}{dy} = \frac{d}{du}\frac{u^2}{2}\frac{d}{dy}\tan^{-1} y = u\frac{1}{1+y^2} = \frac{\tan^{-1} y}{1+y^2}$

Chapter 4 Solutions

① $u = e^x, \dfrac{du}{dx} = e^x, du = e^x dx$

$\displaystyle\int e^x \sinh(e^x)\, dx = \int \sinh u\, du = \cosh u + c = \cosh(e^x) + c$

Check the answer: $f = \cosh u + c = \cosh(e^x) + c, u = e^x$

$\dfrac{df}{dx} = \dfrac{df}{du}\dfrac{du}{dx} = \dfrac{d}{du}(\cosh u + c)\dfrac{d}{dx}e^x = (\sinh u)e^x = e^x \sinh(e^x)$

Note: $\dfrac{d}{dx}\cosh x = \sinh x$ has a different sign than $\dfrac{d}{dx}\cos x = -\sin x$.

Note: The **chain rule** is used to check the answers for this chapter, much like it was used in the examples and in the previous chapters.

② $u = 1 + \ln t, \dfrac{du}{dt} = \dfrac{1}{t}, du = \dfrac{dt}{t}$

$\displaystyle\int \dfrac{\sqrt{1 + \ln t}}{t}\, dt = \int \sqrt{u}\, du = \int u^{1/2}\, du = \dfrac{u^{\frac{1}{2}+1}}{\frac{1}{2}+1} + c = \dfrac{u^{3/2}}{3/2} + c$

$= \dfrac{2u^{3/2}}{3} + c = \dfrac{2(1 + \ln t)^{3/2}}{3} + c$ Alternate answer: $\dfrac{2(1 + \ln t)\sqrt{1 + \ln t}}{3} + c$

Check the answer: $f = \dfrac{2u^{3/2}}{3} + c = \dfrac{2(1+\ln t)^{3/2}}{3} + c, u = 1 + \ln t$

$\dfrac{df}{dt} = \dfrac{df}{du}\dfrac{du}{dt} = \dfrac{d}{du}\left(\dfrac{2u^{3/2}}{3} + c\right)\dfrac{d}{dt}(1 + \ln t) = (u^{1/2})\left(\dfrac{1}{t}\right) = \dfrac{\sqrt{u}}{t} = \dfrac{\sqrt{1 + \ln t}}{t}$

③ $u = 5^w - 4, \dfrac{du}{dw} = 5^w \ln 5, du = 5^w \ln 5\, dw, \dfrac{du}{\ln 5} = 5^w dw$

At $w = 1, u = 5^1 - 4 = 5 - 4 = 1$. At $w = 3, u = 5^3 - 4 = 125 - 4 = 121$.

$\displaystyle\int_{w=1}^{3} \dfrac{5^w}{\sqrt{5^w - 4}}\, dw = \int_{u=1}^{121} \dfrac{du}{\ln 5 \sqrt{u}} = \dfrac{1}{\ln 5}[2\sqrt{u}]_{u=1}^{121} = \dfrac{2}{\ln 5}[\sqrt{u}]_{u=1}^{121}$

$= \dfrac{2}{\ln 5}(\sqrt{121} - \sqrt{1}) = \dfrac{2}{\ln 5}(11 - 1) = \dfrac{2}{\ln 5}(10) = \dfrac{20}{\ln 5} \approx 12.4$

Check the antiderivative: $f = \dfrac{2\sqrt{u}}{\ln 5}, u = 5^w - 4$

$\dfrac{df}{dw} = \dfrac{df}{du}\dfrac{du}{dw} = \dfrac{d}{du}\left(\dfrac{2\sqrt{u}}{\ln 5}\right)\dfrac{d}{dw}(5^w - 4) = \dfrac{1}{\ln 5 \sqrt{u}} 5^w \ln 5 = \dfrac{5^w}{\sqrt{5^w - 4}}$

169

Answers with Full Solutions

4 $\int_{y=1/e}^{1} \ln\left(\frac{1}{y}\right) dy = -\int_{y=1/e}^{1} \ln y \, dy = -[y \ln y - y]_{y=1/e}^{1}$

$= -\left[(1 \ln 1 - 1) - \left(\frac{1}{e}\ln\frac{1}{e} - \frac{1}{e}\right)\right] = -(1 \ln 1 - 1) + \left(\frac{1}{e}\ln\frac{1}{e} - \frac{1}{e}\right)$

$= -[1(0) - 1] + \left(\frac{1}{e}\ln e^{-1} - \frac{1}{e}\right) = -(0 - 1) + \left(-\frac{1}{e}\ln e - \frac{1}{e}\right)$

$= -(-1) + \left[-\frac{1}{e}(1) - \frac{1}{e}\right] = 1 + \left(-\frac{1}{e} - \frac{1}{e}\right) = 1 - \frac{2}{e} = \frac{e-2}{e} \approx 0.264$

Note: We used the identity $\ln\left(\frac{1}{y}\right) = -\ln y$. Similarly, $\ln\left(\frac{1}{e}\right) = -\ln e$.

Check the antiderivative: $\frac{d}{dy}[-(y \ln y - y)] = \frac{d}{dy}(-y \ln y + y)$

$= -y\frac{d}{dy}\ln y - \ln y \frac{d}{dy}y + \frac{d}{dy}y = -y\left(\frac{1}{y}\right) - \ln y(1) + 1 = -\frac{y}{y} - \ln y + 1$

$= -1 - \ln y + 1 = -\ln y = \ln\left(\frac{1}{y}\right)$

5 $\int \coth z \, dz = \int \frac{\cosh z}{\sinh z} dz$

$u = \sinh z, \frac{du}{dz} = \cosh z, du = \cosh z \, dz$

$\int \frac{\cosh z}{\sinh z} dz = \int \frac{du}{u} = \ln|u| + c = \ln|\sinh z| + c$

Check the answer: $f = \ln u + c = \ln(\sinh z) + c, u = \sinh z$

$\frac{df}{dz} = \frac{df}{du}\frac{du}{dz} = \frac{d}{du}\ln u \frac{d}{dz}\sinh z = \frac{1}{u}\cosh z = \frac{\cosh z}{\sinh z} = \coth z$

6 $\tanh^{-1} y = \frac{1}{2}\ln\left(\frac{1+y}{1-y}\right)$ provided that $|y| < 1$

$\int \tanh^{-1} y \, dy = \frac{1}{2}\int \ln\left(\frac{1+y}{1-y}\right) dy = \frac{1}{2}\int \ln(1+y) \, dy - \frac{1}{2}\int \ln(1-y) \, dy$

Note: We used the difference of logs formula: $\ln\left(\frac{p}{q}\right) = \ln p - \ln q$.

$u = 1 + y, du = dy, v = 1 - y, dv = -dy$

$\frac{1}{2}\int \ln(1+y) \, dy - \frac{1}{2}\int \ln(1-y) \, dy = \frac{1}{2}\int \ln u \, du + \frac{1}{2}\int \ln v \, dv$

$$= \frac{1}{2}(u\ln u - u) + \frac{1}{2}(v\ln v - v) + c = \frac{u\ln u + v\ln v - (u+v)}{2} + c$$

$$= \frac{(1+y)\ln(1+y) + (1-y)\ln(1-y) - (1+y+1-y)}{2} + c$$

$$= \frac{\ln(1+y) + y\ln(1+y) + \ln(1-y) - y\ln(1-y) - 2}{2} + c$$

$$= \frac{\ln[(1+y)(1-y)] + y\ln\left(\frac{1+y}{1-y}\right)}{2} - \frac{2}{2} + c = \frac{\ln(1-y^2) + y\ln\left(\frac{1+y}{1-y}\right)}{2} + c_2$$

$$= \frac{\ln(1-y^2)}{2} + \frac{y\ln\left(\frac{1+y}{1-y}\right)}{2} + c_2 = \frac{\ln(1-y^2)}{2} + y\tanh^{-1} y + c_2$$

Notes: $\tanh^{-1} y$ is the **inverse** hyperbolic tangent function. It is not a reciprocal; it does **not** mean one divided by $\tanh y$. Rather, $\tanh^{-1} y$ asks, what can you take the hyperbolic tangent of and obtain y as a result?

We used the identities $\ln(pq) = \ln p + \ln q$ and $\ln\left(\frac{p}{q}\right) = \ln p - \ln q$. In the last step, we again used $\tanh^{-1} y = \frac{1}{2}\ln\left(\frac{1+y}{1-y}\right)$.

Since $-\frac{2}{2} = -1$ is just a constant, $-\frac{2}{2} + c$ makes a new constant. We called it c_2 just to distinguish it from the first constant that we introduced: $-\frac{2}{2} + c = c_2$.

Alternate answers: There is more than one way to express this answer; some of the alternates appear above. However, the form $y\tanh^{-1} y + \frac{\ln(1-y^2)}{2}$ + a constant is often preferred by books and instructors because it is fairly concise.

⑦ $\int_{t=0}^{1} \sinh t \, dt = \int_{t=0}^{1} \frac{e^t - e^{-t}}{2} dt = \left[\frac{e^t - (-e^{-t})}{2}\right]_{t=0}^{1} = \left[\frac{e^t + e^{-t}}{2}\right]_{t=0}^{1} = [\cosh t]_{t=0}^{1}$

$= \cosh 1 - \cosh 0 = \cosh 1 - 1 = -1 + \cosh 1 \approx -1 + 1.543 \approx 0.543$

Check the antiderivative: $\frac{d}{dt}\cosh t = \sinh t$ (Note that this hyperbolic derivative has a different sign compared to the ordinary trig derivative.)

⑧ $\int_{x=0}^{1} \cosh^2 x \, dx = \int_{x=0}^{1} \left(\frac{e^x + e^{-x}}{2}\right)^2 dx = \int_{x=0}^{1} \frac{e^{2x} + 2 + e^{-2x}}{4} dx$

Notes: $e^x e^x = e^{x+x} = e^{2x}$, $e^x e^{-x} = e^{x-x} = e^0 = 1$, $e^{-x}e^{-x} = e^{-x-x} = e^{-2x}$

Answers with Full Solutions

$$\int_{x=0}^{1} \frac{e^{2x} + 2 + e^{-2x}}{4} dx = \frac{1}{4} \int_{x=0}^{1} (e^{2x} + 2 + e^{-2x}) dx$$

$$= \frac{1}{4} \int_{x=0}^{1} e^{2x} dx + \frac{1}{4} \int_{x=0}^{1} 2 dx + \frac{1}{4} \int_{x=0}^{1} e^{-2x} dx = \frac{1}{4} \left[\frac{e^{2x}}{2}\right]_{x=0}^{1} + \frac{1}{4}[2x]_{x=0}^{1} + \frac{1}{4}\left[-\frac{e^{-2x}}{2}\right]_{x=0}^{1}$$

$$= \frac{1}{8}(e^2 - e^0) + \frac{1}{2}(1 - 0) - \frac{1}{8}(e^{-2} - e^0) = \frac{e^2}{8} - \frac{1}{8} + \frac{1}{2} - 0 - \frac{e^{-2}}{8} + \frac{1}{8}$$

$$= \frac{e^2 - e^{-2}}{8} + \frac{1}{2} \approx 1.41 \quad \text{Note: } \frac{e^2 - e^{-2}}{8} + \frac{1}{2} = \frac{\sinh 2}{4} + \frac{1}{2} = \frac{\sinh 2 + 2}{4} = \frac{2 + \sinh 2}{4}$$

Check the antiderivative: $f = \frac{e^{2x}}{4(2)} + \frac{2x}{4} - \frac{e^{-2x}}{4(2)} = \frac{e^{2x}}{8} + \frac{x}{2} - \frac{e^{-2x}}{8}$. Apply the chain rule.

$$\frac{df}{dx} = \frac{d}{dx}\left(\frac{e^{2x}}{8} + \frac{x}{2} - \frac{e^{-2x}}{8}\right) = \frac{e^{2x}}{8}(2) + \frac{1}{2} - \frac{e^{-2x}}{8}(-2) = \frac{e^{2x}}{4} + \frac{2}{4} + \frac{e^{-2x}}{4}$$

$$= \frac{(e^x + e^{-x})^2}{4} = \left(\frac{e^x + e^{-x}}{2}\right)^2 = \cosh^2 x$$

9 $\int \frac{2\tanh z}{1 + \tanh^2 z} dz = \int \tanh(2z) dz \quad$ Note: $\tanh(2z) = \frac{2\tanh z}{1 + \tanh^2 z}$

$$u = 2z, \frac{du}{dz} = 2, du = 2dz, \frac{du}{2} = dz$$

$$\int \tanh(2z) dz = \int \frac{\tanh u}{2} du = \frac{1}{2} \int \tanh u \, du = \frac{1}{2} \ln(\cosh u) + c$$

$$= \frac{1}{2} \ln[\cosh(2z)] + c$$

Note: Since $\cosh u > 0$, no absolute value symbols are needed in the natural log.

Check the answer: $f = \frac{1}{2}\ln w + c = \frac{1}{2}\ln(\cosh u) + c = \frac{1}{2}\ln[\cosh(2z)] + c$,

$w = \cosh u, u = 2z$

$$\frac{df}{dz} = \frac{df}{dw}\frac{dw}{du}\frac{du}{dz} = \frac{d}{dw}\frac{1}{2}\ln w \frac{d}{du}\cosh u \frac{d}{dz} 2z = \frac{1}{2w}\sinh u \,(2) = \frac{\sinh u}{w}$$

$$= \frac{\sinh(2z)}{\cosh u} = \frac{\sinh(2z)}{\cosh(2z)} = \tanh(2z)$$

10 $\int (\cosh y + \sinh y)^3 \, dy = \int e^{3y} dy = \frac{e^{3y}}{3} + c = \frac{\cosh(3y)}{3} + \frac{\sinh(3y)}{3} + c$

Alternate answers: $\frac{1}{3}(\cosh y + \sinh y)^3 + c$

$$= \frac{\cosh^3 y}{3} + \cosh^2 y \sinh y + \cosh y \sinh^2 y + \frac{\sinh^3 y}{3} + c$$

Note: $(\cosh y + \sinh y)^3 = e^{3y} = \cosh(3y) + \sinh(3y)$

Check the answer: $\frac{d}{dy}\left(\frac{e^{3y}}{3} + c\right) = 3\left(\frac{e^{3y}}{3}\right) = e^{3y} = (\cosh y + \sinh y)^3$

⑪ $u = \tanh x, \frac{du}{dx} = \text{sech}^2 x, du = \text{sech}^2 x\, dx$

At $x = 0, u = \tanh 0 = 0$. At $x = 1, u = \tanh 1 \approx 0.7616$.

$$\int_{x=0}^{1} \tanh x\, \text{sech}^2 x\, dx \approx \int_{u=0}^{0.7616} u\, du = \left[\frac{u^2}{2}\right]_0^{0.7616} = \frac{0.7616^2}{2} - \frac{0^2}{2} \approx 0.290$$

Check the antiderivative: $f = \frac{u^2}{2} = \frac{\tanh^2 x}{2}, u = \tanh x$

$\frac{df}{dx} = \frac{df}{du}\frac{du}{dx} = \frac{d}{du}\frac{u^2}{2}\frac{d}{dx}\tanh x = u\, \text{sech}^2 x = \tanh x\, \text{sech}^2 x$

⑫ $\int_{w=0}^{1} \frac{1+\tanh w}{1-\tanh w} dw = \int_{w=0}^{1} e^{2w} dw$ Note: $e^{2w} = \frac{1+\tanh w}{1-\tanh w}$

$u = 2w, \frac{du}{dw} = 2, du = 2dw, \frac{du}{2} = dw$

At $w = 0, u = 2(0) = 0$. At $w = 1, u = 2(1) = 2$.

$$\int_{w=0}^{1} e^{2w} dw = \int_{u=0}^{2} \frac{e^u}{2} du = \frac{1}{2}\int_{u=0}^{2} e^u du = \frac{1}{2}[e^u]_{u=0}^{2} = \frac{e^2 - e^0}{2} = \frac{e^2 - 1}{2} \approx 3.19$$

Check the antiderivative: $f = \frac{e^u}{2} = \frac{e^{2w}}{2}, u = 2w$

$\frac{df}{dw} = \frac{df}{du}\frac{du}{dw} = \frac{d}{du}\frac{e^u}{2}\frac{d}{dw}2w = \frac{e^u}{2}(2) = e^u = e^{2w} = \frac{1+\tanh w}{1-\tanh w}$

Chapter 5 Solutions

❶ $x^2 + 12x + 21$; $a = 1, b = 12, c = 21$

$$\left(x\sqrt{a} + \frac{b}{2\sqrt{a}}\right)^2 + \left(c - \frac{b^2}{4a}\right) = \left(x\sqrt{1} + \frac{12}{2\sqrt{1}}\right)^2 + \left[21 - \frac{12^2}{4(1)}\right]$$

$$= (x + 6)^2 + \left(21 - \frac{144}{4}\right) = (x + 6)^2 + (21 - 36) = (x + 6)^2 - 15$$

Check: $(x + 6)^2 - 15 = x^2 + 2(x)(6) + 36 - 15 = x^2 + 12 + 21$

❷ $4y^2 + 20y + 42$; $a = 4, b = 20, c = 42$

$$\left(y\sqrt{4} + \frac{20}{2\sqrt{4}}\right)^2 + \left[42 - \frac{20^2}{4(4)}\right] = \left[2y + \frac{20}{2(2)}\right]^2 + \left(42 - \frac{400}{16}\right)$$

$$= (2y + 5)^2 + (42 - 25) = (2y + 5)^2 + 17$$

Check: $(2y + 5)^2 + 17 = 4y^2 + 2(2y)(5) + 25 + 17 = 4y + 20y + 42$

❸ $25w^2 - 30w - 6$; $a = 25, b = -30, c = -6$

$$\left(w\sqrt{25} + \frac{-30}{2\sqrt{25}}\right)^2 + \left[-6 - \frac{(-30)^2}{4(25)}\right] = \left[5w - \frac{30}{2(5)}\right]^2 + \left(-6 - \frac{900}{100}\right)$$

$$= (5w - 3)^2 + (-6 - 9) = (5w - 3)^2 - 15$$

Check: $(5w - 3)^2 - 15 = 25w^2 + 2(5w)(-3) + 9 - 15 = 25w^2 - 30w - 6$

❹ $75 - 96t - 36t^2 = (-1)(-75 + 96t + 36t^2) = (-1)(36t^2 + 96t - 75)$

$a = 36, b = 96, c = -75$ (see Example 3)

$$(-1)\left\{\left(t\sqrt{36} + \frac{96}{2\sqrt{36}}\right)^2 + \left[-75 - \frac{96^2}{4(36)}\right]\right\} = (-1)\left\{\left[6t + \frac{96}{2(6)}\right]^2 + \left(-75 - \frac{9216}{144}\right)\right\}$$

$$= (-1)[(6t + 8)^2 + (-75 - 64)] = (-1)[(6t + 8)^2 - 139] = 139 - (6t + 8)^2$$

Alternate answer: $139 - 4(3t + 4)^2$

Check: $139 - (6t + 8)^2 = 139 - [36t^2 + 2(6t)(8) + 64]$

$= 139 - (36t^2 + 96t + 64) = 139 - 36t^2 - 96t - 64 = -36t^2 - 96t + 75$

❺ $81u^2 + 54u - 1$; $a = 81, b = 54, c = -1$

$$\left(u\sqrt{81} + \frac{54}{2\sqrt{81}}\right)^2 + \left[-1 - \frac{54^2}{4(81)}\right] = \left[9u + \frac{54}{2(9)}\right]^2 + \left(-1 - \frac{2916}{324}\right)$$

$= (9y + 3)^2 + (-1 - 9) = (9u + 3)^2 - 10$ Alternate answer: $9(3u + 1)^2 - 10$

Check: $(9u + 3)^2 - 10 = 81u^2 + 2(9u)(3) + 9 - 10 = 81u^2 + 54u - 1$

❻ $16x^2 - 32x + 11$; $a = 16, b = -32, c = 11$

$$\left(x\sqrt{16} + \frac{-32}{2\sqrt{16}}\right)^2 + \left[11 - \frac{(-32)^2}{4(16)}\right] = \left[4x - \frac{32}{2(4)}\right]^2 + \left(11 - \frac{1024}{64}\right)$$

$= (4x - 4)^2 + (11 - 16) = (4x - 4)^2 - 5$ Alternate answer: $16(x - 1)^2 - 5$

Check: $(4x - 4)^2 - 5 = 16x^2 + 2(4x)(-4) + 16 - 5 = 16x^2 - 32x + 11$

❼ $50 - 42z^2 + 9z^4 = 9z^4 - 42z^2 + 50$

Let $u = z^2$ to get $9u^2 - 42u + 50$; $a = 9, b = -42, c = 50$

$$\left(u\sqrt{9} + \frac{-42}{2\sqrt{9}}\right)^2 + \left[50 - \frac{(-42)^2}{4(9)}\right] = \left[3u - \frac{42}{2(3)}\right]^2 + \left(50 - \frac{1764}{36}\right)$$

$= (3u - 7)^2 + (50 - 49) = (3u - 7)^2 + 1 = (3z^2 - 7)^2 + 1$

Check: $(3z^2 - 7)^2 + 1 = 9z^4 + 2(3z^2)(-7) + 49 + 1 = 9z^4 - 42z^2 + 50$

❽ $144y^2 - 48y$; $a = 144, b = -48, c = 0$

$$\left(y\sqrt{144} + \frac{-48}{2\sqrt{144}}\right)^2 + \left[0 - \frac{(-48)^2}{4(144)}\right] = \left[12y - \frac{48}{2(12)}\right]^2 - \frac{2304}{576} = (12y - 2)^2 - 4$$

Alternate answers: $4(6y - 1)^2 - 4 = 4[(6y - 1)^2 - 1]$

Check: $(12y - 2)^2 - 4 = 144y^2 + 2(12y)(-2) + 4 - 4 = 144y^2 - 48y$

Alternate answer/solution: If you want to be 'cute,' there is another way to complete the 'square.' **Factor** out a y, and then add a **higher power** to the polynomial. This is not easy for most students to figure out on their own (but it would be a good challenge for advanced students; if that's you, avoid looking at the solution below). If you're curious, see if you can follow the steps of the alternate solution below. The logic is basically 'backwards' compared to usual. Here, we're taking the square root of the constant, and completing the square to figure out what the quadratic term should be; our usual formula won't work (in the usual way, at least). To start out, the square root of 48 is equal to $4\sqrt{3}$ (since $48 = 16 \times 3$).

$144y^2 - 48y = -y(-144y + 48) = -y(48 - 144y) = -y\left[\left(4\sqrt{3}\right)^2 - 48y\sqrt{3}\sqrt{3}\right]$

$= -y\left[\left(4\sqrt{3}\right)^2 - 48y\sqrt{3}\sqrt{3} + \left(6y\sqrt{3}\right)^2 - \left(6y\sqrt{3}\right)^2\right]$

$= -y\left[\left(4\sqrt{3} - 6y\sqrt{3}\right)^2 - 108y^2\right] = -y\left[\left(6y\sqrt{3} - 4\sqrt{3}\right)^2 - 108y^2\right]$

$= -y[3(6y - 4)^2 - 108y^2]$

Alternate answers: $-3y[(6y - 4)^2 - 36y^2] = -12y[(3y - 2)^2 - 9y^2]$

Answers with Full Solutions

Check: $-3y[(6y-4)^2 - 36y^2] = -3y[36y^2 + 2(6y)(-4) + 16 - 36y^2]$
$= -3y(-48y + 16) = 144y^2 - 48y$

❾ $121 + 140t - 49t^2 = (-1)(-121 - 140t + 49t^2) = (-1)(49t^2 - 140t - 121)$
$a = 49, b = -140, c = -121$ (see Example 3)

$$(-1)\left\{\left(t\sqrt{49} + \frac{-140}{2\sqrt{49}}\right)^2 + \left[-121 - \frac{(-140)^2}{4(49)}\right]\right\}$$

$$= (-1)\left\{\left[7t - \frac{140}{2(7)}\right]^2 + \left(-121 - \frac{19{,}600}{196}\right)\right\}$$

$= (-1)[(7t-10)^2 + (-121 - 100)] = (-1)[(7t-10)^2 - 221] = 221 - (7t-10)^2$

Check: $221 - (7t-10)^2 = 221 - [49t^2 + 2(7t)(-10) + 100]$
$= 221 - (49t^2 - 140t + 100) = 221 - 49t^2 + 140t - 100 = -49t^2 + 140t + 121$

❿ $12w^2 - 8w\sqrt{3} - 2; \; a = 12, b = -8\sqrt{3}, c = -2$

$$\left(w\sqrt{12} + \frac{-8\sqrt{3}}{2\sqrt{12}}\right)^2 + \left[-2 - \frac{(-8\sqrt{3})^2}{4(12)}\right] = \left[w2\sqrt{3} - \frac{8\sqrt{3}}{2(2\sqrt{3})}\right]^2 + \left[-2 - \frac{(64)(3)}{48}\right]$$

$= (2w\sqrt{3} - 2)^2 + (-2 - 4) = (2w\sqrt{3} - 2)^2 - 6$

Notes: $\sqrt{12} = \sqrt{(4)(3)} = \sqrt{4}\sqrt{3} = 2\sqrt{3}$ and $(-8\sqrt{3})^2 = (-8)^2(\sqrt{3})^2 = 64(3)$.

Check: $(2w\sqrt{3} - 2)^2 - 6 = (2)^2 w^2 (\sqrt{3})^2 + 2(2w\sqrt{3})(-2) + 4 - 6$
$= 4w^2(3) - 8w\sqrt{3} - 2 = 12w^2 - 8w\sqrt{3} - 2$

⓫ $9x + 24\sqrt{x} + 2$

Let $u = \sqrt{x}$ to get $9u^2 + 24u + 2$; $a = 9, b = 24, c = 2$

$$\left(u\sqrt{9} + \frac{24}{2\sqrt{9}}\right)^2 + \left[2 - \frac{24^2}{4(9)}\right] = \left[3u + \frac{24}{2(3)}\right]^2 + \left(2 - \frac{576}{36}\right)$$

$= (3u + 4)^2 + (2 - 16) = (3u + 4)^2 - 14 = (3\sqrt{x} + 4)^2 - 14$

Check: $(3\sqrt{x} + 4)^2 - 14 = 9x + 2(3\sqrt{x})(4) + 16 - 14 = 9x + 24\sqrt{x} + 2$

⓬ $4\cos^2\theta + 4\cos\theta + 5; \; a = 4, b = 4, c = 5$

$$\left(\cos\theta\sqrt{4} + \frac{4}{2\sqrt{4}}\right)^2 + \left[5 - \frac{4^2}{4(4)}\right] = \left[2\cos\theta + \frac{4}{2(2)}\right]^2 + \left(5 - \frac{16}{16}\right)$$

$= (2\cos\theta + 1)^2 + (5 - 1) = (2\cos\theta + 1)^2 + 4$

Check: $(2\cos\theta + 1)^2 + 4 = 4\cos^2\theta + 2(2\cos\theta)(1) + 1 + 4 = 4\cos^2\theta + 4\cos\theta + 5$

❸ $x^2 - 10x + 29 = (x-5)^2 + 4$

Check: $(x-5)^2 + 4 = x^2 + 2(x)(-5) + 25 + 4 = x^2 - 10x + 29$

$$\int \frac{dx}{\sqrt{x^2 - 10x + 29}} = \int \frac{dx}{\sqrt{(x-5)^2 + 4}}$$

$u = x - 5, \dfrac{du}{dx} = 1, du = dx$

$$\int \frac{dx}{\sqrt{(x-5)^2 + 4}} = \int \frac{du}{\sqrt{u^2 + 4}}$$

$u = 2\tan\theta, du = 2\sec^2\theta\, d\theta$

$$\int \frac{du}{\sqrt{u^2+4}} = \int \frac{2\sec^2\theta}{\sqrt{4\tan^2\theta + 4}} d\theta = \int \frac{2\sec^2\theta}{\sqrt{4\sec^2\theta}} d\theta = \int \frac{\sec^2\theta}{\sec\theta} d\theta = \int \sec\theta\, d\theta$$

$$= \ln|\sec\theta + \tan\theta| + c = \ln\left|\sec\left(\tan^{-1}\frac{u}{2}\right) + \tan\left(\tan^{-1}\frac{u}{2}\right)\right| + c$$

$$= \ln\left|\sqrt{1 + \frac{u^2}{4}} + \frac{u}{2}\right| + c = \ln\left|\frac{u}{2} + \sqrt{1 + \frac{u^2}{4}}\right| + c = \ln\left|\frac{x-5}{2} + \sqrt{1 + \frac{(x-5)^2}{4}}\right| + c$$

Alternate answer: $\ln\left|2x - 10 + 2\sqrt{4 + (x-5)^2}\right| + c$. See the note in Chapter 6, Ex. 1.

Note: See the footnotes in Chapter 3 regarding inverse trig identities.

Check the answer: Apply the chain rule.

$$\frac{d}{dx}\left[\ln\left|\frac{x-5}{2} + \sqrt{1 + \frac{(x-5)^2}{4}}\right| + c\right] = \frac{1}{\frac{x-5}{2} + \sqrt{1 + \frac{(x-5)^2}{4}}}\left[\frac{1}{2} + \frac{\frac{2(x-5)}{4}(1)}{2\sqrt{1 + \frac{(x-5)^2}{4}}}\right]$$

$$= \frac{1}{\frac{x-5}{2} + \sqrt{1 + \frac{(x-5)^2}{4}}}\left[\frac{\sqrt{1 + \frac{(x-5)^2}{4}}}{2\sqrt{1 + \frac{(x-5)^2}{4}}} + \frac{\frac{x-5}{2}}{2\sqrt{1 + \frac{(x-5)^2}{4}}}\right]$$

$$= \frac{1}{\frac{x-5}{2} + \sqrt{1 + \frac{(x-5)^2}{4}}} \cdot \frac{\frac{x-5}{2} + \sqrt{1 + \frac{(x-5)^2}{4}}}{2\sqrt{1 + \frac{(x-5)^2}{4}}} = \frac{1}{2\sqrt{1 + \frac{(x-5)^2}{4}}}$$

$$= \frac{1}{\sqrt{4}\sqrt{1 + \frac{(x-5)^2}{4}}} = \frac{1}{\sqrt{4 + (x-5)^2}} = \frac{1}{\sqrt{(x-5)^2 + 4}} = \frac{1}{\sqrt{x^2 - 10x + 29}}$$

Answers with Full Solutions

⑭ $25y^2 - 10y - 8 = (5y - 1)^2 - 9$

Check: $(5y - 1)^2 - 9 = 25y^2 + 2(5y)(-1) + 1 - 9 = 25y^2 - 10y - 8$

$$\int \frac{dy}{25y^2 - 10y - 8} = \int \frac{dy}{(5y - 1)^2 - 9}$$

$u = 5y - 1, du = 5dy, \dfrac{du}{5} = dy$

$$\int \frac{dy}{(5y - 1)^2 - 9} = \frac{1}{5}\int \frac{du}{u^2 - 9}$$

$u = 3\sec\theta, du = 3\sec\theta\tan\theta\, d\theta$

$$\frac{1}{5}\int \frac{du}{u^2 - 9} = \frac{1}{5}\int \frac{3\sec\theta\tan\theta}{9\sec^2\theta - 9}d\theta = \frac{1}{15}\int \frac{\sec\theta\tan\theta}{\tan^2\theta}d\theta = \frac{1}{15}\int \frac{\sec\theta}{\tan\theta}d\theta$$

$$= \frac{1}{15}\int \frac{1}{\cos\theta}\frac{\cos\theta}{\sin\theta}d\theta = \frac{1}{15}\int \csc\theta\, d\theta = \frac{1}{15}\ln|\csc\theta - \cot\theta| + c$$

$$= \frac{1}{15}\ln|\csc\theta - \cot\theta| + c = \frac{1}{15}\ln\left|\csc\left(\sec^{-1}\frac{u}{3}\right) - \cot\left(\sec^{-1}\frac{u}{3}\right)\right| + c$$

$$= \frac{1}{15}\ln\left|\frac{u/3}{\sqrt{u^2/9 - 1}} - \frac{1}{\sqrt{u^2/9 - 1}}\right| + c = \frac{1}{15}\ln\left|\frac{u}{3\sqrt{u^2/9 - 1}} - \frac{3}{3\sqrt{u^2/9 - 1}}\right| + c$$

$$= \frac{1}{15}\ln\left|\frac{u}{\sqrt{u^2 - 9}} - \frac{3}{\sqrt{u^2 - 9}}\right| + c = \frac{1}{15}\ln\left|\frac{u - 3}{\sqrt{u^2 - 9}}\right| + c = \frac{1}{15}\ln\left|\frac{u - 3}{\sqrt{(u + 3)(u - 3)}}\right| + c$$

$$= \frac{1}{15}\ln\sqrt{\left|\frac{u - 3}{u + 3}\right|} + c = \frac{1}{30}\ln\left|\frac{u - 3}{u + 3}\right| + c = \frac{1}{30}\ln\left|\frac{5y - 4}{5y + 2}\right| + c$$

Notes: We used the rule $\ln\sqrt{z} = \ln z^{1/2} = \frac{1}{2}\ln z$.

Thought: Did you notice that $25y^2 - 10y - 8$ factors as $(5y - 4)(5y + 2)$? Another way to solve this problem is to use the method of partial fractions (Chapter 6).

Check the answer: Note that $\frac{1}{30}\ln\left(\frac{5y-4}{5y+2}\right) = \frac{1}{30}\ln(5y - 4) - \frac{1}{30}\ln(5y + 2)$. Apply the chain rule.

$$\frac{d}{dy}\left[\frac{1}{30}\ln(5y - 4) - \frac{1}{30}\ln(5y + 2) + c\right] = \frac{1}{30(5y - 4)}\frac{d}{dy}(5y - 4) - \frac{1}{30(5y + 2)}\frac{d}{dy}(5y + 2)$$

$$= \frac{5}{30(5y - 4)} - \frac{5}{30(5y + 2)} = \frac{1}{6(5y - 4)} - \frac{1}{6(5y + 2)} = \frac{5y + 2 - (5y - 4)}{6(5y - 4)(5y + 2)}$$

$$= \frac{5y + 2 - 5y + 4}{6(25y^2 + 10y - 20y - 8)} = \frac{6}{6(25y^2 - 10y - 8)} = \frac{1}{25y^2 - 10y - 8}$$

⑮ $8 + 6p - 9p^2 = 9 - (1-3p)^2 = 9 - (3p-1)^2$

Note: $(1-3p)^2 = [(-1)(3p-1)]^2 = (-1)^2(3p-1)^2 = (3p-1)^2$

Check: $9 - (3p-1)^2 = 9 - [9p^2 + 2(3p)(-1) + 1] = 9 - (9p^2 - 6p + 1)$
$= 9 - 9p^2 + 6p - 1 = -9p^2 + 6p + 8$

$$\int \frac{dp}{\sqrt{8 + 6p - 9p^2}} = \int \frac{dp}{\sqrt{9 - (3p-1)^2}}$$

$u = 3p - 1, du = 3dp, \dfrac{du}{3} = dp$

$$\int \frac{dp}{\sqrt{9 - (3p-1)^2}} = \frac{1}{3}\int \frac{du}{\sqrt{9 - u^2}}$$

$u = 3\sin\theta, du = 3\cos\theta\, d\theta$

$$\frac{1}{3}\int \frac{du}{\sqrt{9 - u^2}} = \frac{1}{3}\int \frac{3\cos\theta}{\sqrt{9 - 9\sin^2\theta}} d\theta = \int \frac{\cos\theta}{\sqrt{9\cos^2\theta}} d\theta = \frac{1}{3}\int \frac{\cos\theta}{\cos\theta} d\theta$$

$$= \frac{1}{3}\int d\theta = \frac{\theta}{3} + c = \frac{1}{3}\sin^{-1}\left(\frac{u}{3}\right) + c = \frac{1}{3}\sin^{-1}\left(\frac{3p-1}{3}\right) + c = \frac{1}{3}\sin^{-1}\left(p - \frac{1}{3}\right) + c$$

Check the answer: Apply the chain rule. Recall from first-semester calculus that
$\dfrac{d}{dp}\sin^{-1}w = \dfrac{1}{\sqrt{1-w^2}}\dfrac{dw}{dp}$ provided that $-\dfrac{\pi}{2} < \sin^{-1}w < \dfrac{\pi}{2}$.

$$\frac{d}{dp}\left[\frac{1}{3}\sin^{-1}\left(\frac{3p-1}{3}\right) + c\right] = \frac{1}{3}\frac{1}{\sqrt{1 - \left(\frac{3p-1}{3}\right)^2}}\frac{d}{dp}\left(\frac{3p-1}{3}\right) = \frac{1}{3}\frac{3/3}{\sqrt{1 - \left(\frac{3p-1}{3}\right)^2}}$$

$$= \frac{1}{\sqrt{9}\sqrt{1 - \frac{9p^2 - 6p + 1}{9}}} = \frac{1}{\sqrt{9 - (9p^2 - 6p + 1)}} = \frac{1}{\sqrt{9 - 9p^2 + 6p - 1}} = \frac{1}{\sqrt{8 + 6p - 9p^2}}$$

⑯ $16w^2 + 24w = (4w + 3)^2 - 9$

Check: $(4w+3)^2 - 9 = 16w^2 + 2(4w)(3) + 9 - 9 = 16w^2 + 24w$

$$\int \frac{dw}{\sqrt{16w^2 + 24w}} = \int \frac{dw}{\sqrt{(4w+3)^2 - 9}}$$

$u = 4w + 3, du = 4dw, \dfrac{du}{4} = dw$

$$\int \frac{dw}{\sqrt{(4w+3)^2 - 9}} dw = \frac{1}{4}\int \frac{du}{\sqrt{u^2 - 9}}$$

$u = 3\sec\theta, du = 3\sec\theta\tan\theta\, d\theta$

179

Answers with Full Solutions

$$\frac{1}{4}\int \frac{du}{\sqrt{u^2-9}} = \frac{1}{4}\int \frac{3\sec\theta\tan\theta}{\sqrt{9\sec^2\theta - 9}}d\theta$$

$$= \frac{1}{4}\int \frac{3\sec\theta\tan\theta}{\sqrt{9\tan^2\theta}}d\theta = \frac{1}{4}\int \frac{3\sec\theta\tan\theta}{3\tan\theta}d\theta = \frac{1}{4}\int \sec\theta\, d\theta$$

$$= \frac{1}{4}\ln|\sec\theta + \tan\theta| + c = \frac{1}{4}\ln\left|\sec\left(\sec^{-1}\frac{u}{3}\right) + \tan\left(\sec^{-1}\frac{u}{3}\right)\right| + c$$

$$= \frac{1}{4}\ln\left|\frac{u}{3} + \sqrt{\frac{u^2}{9} - 1}\right| + c = \frac{1}{4}\ln\left|\frac{4w+3}{3} + \sqrt{\frac{(4w+3)^2}{9} - 1}\right| + c$$

$$= \frac{1}{4}\ln\left|\frac{4w+3}{3} + \sqrt{\frac{16w^2 + 24w + 9}{9} - \frac{9}{9}}\right| + c = \frac{1}{4}\ln\left|\frac{4w+3}{3} + \sqrt{\frac{16w^2 + 24w}{9}}\right| + c$$

Alternate answer: $\frac{1}{4}\ln\left|4w + 3 + \sqrt{16w^2 + 24w}\right| + c$. See the note in Chapter 6, Ex. 1.

Note: See the footnotes in Chapter 3 regarding inverse trig identities.

Check the answer: Apply the chain rule.

$$\frac{d}{dw}\left[\frac{1}{4}\ln\left|\frac{4w+3}{3} + \sqrt{\frac{16w^2+24w}{9}}\right| + c\right] = \frac{1/4}{\frac{4w+3}{3} + \sqrt{\frac{16w^2+24w}{9}}}\left[\frac{4}{3} + \frac{\frac{32w+24}{9}}{2\sqrt{\frac{16w^2+24w}{9}}}\right]$$

Note: $\dfrac{\frac{32w+24}{9}}{2\sqrt{\frac{16w^2+24w}{9}}} = \dfrac{\frac{16w+12}{9}}{\sqrt{\frac{16w^2+24w}{9}}} = \dfrac{\frac{16w+12}{3}}{3\sqrt{\frac{16w^2+24w}{9}}} = \dfrac{\frac{4(4w+3)}{3}}{3\sqrt{\frac{16w^2+24w}{9}}}$. (This is a good fractions refresher.)

$$= \frac{1/4}{\frac{4w+3}{3} + \sqrt{\frac{16w^2+24w}{9}}}\left[\frac{4\sqrt{\frac{16w^2+24w}{9}}}{3\sqrt{\frac{16w^2+24w}{9}}} + \frac{\frac{4(4w+3)}{3}}{3\sqrt{\frac{16w^2+24w}{9}}}\right]$$

$$= \frac{1/4}{\frac{4w+3}{3} + \sqrt{\frac{16w^2+24w}{9}}}\left[\frac{\sqrt{\frac{16w^2+24w}{9}} + \frac{(4w+3)}{3}}{\sqrt{\frac{16w^2+24w}{9}}}\right]\left(\frac{4}{3}\right) \quad \left(\text{We factored out } \frac{4}{3}\right)$$

$$= \frac{\frac{1}{4}}{\sqrt{\frac{16w^2+24w}{9}}}\left(\frac{4}{3}\right) = \frac{1}{3\sqrt{\frac{(4w+3)^2}{9} - 1}} = \frac{1}{\sqrt{(4w+3)^2 - 9}}$$

$$= \frac{1}{\sqrt{16w^2 + 24w + 9 - 9}} = \frac{1}{\sqrt{16w^2 + 24w}}$$

180

⑰ $4x^4 + 12x^2 + 25 = (2x^2 + 3)^2 + 16$

Check: $(2x^2 + 3)^2 + 16 = 4x^4 + 2(2x^2)(3) + 9 + 16 = 4x^4 + 12x^2 + 25$

$$\int \frac{x}{4x^4 + 12x^2 + 25} dx = \int \frac{x}{(2x^2 + 3)^2 + 16} dx$$

$u = 2x^2 + 3, du = 4xdx, \dfrac{du}{4} = xdx$

$$\int \frac{x}{(2x^2 + 3)^2 + 16} dx = \frac{1}{4}\int \frac{du}{u^2 + 16}$$

$u = 4\tan\theta, du = 4\sec^2\theta \, d\theta$

$$\frac{1}{4}\int \frac{du}{u^2 + 16} = \frac{1}{4}\int \frac{4\sec^2\theta}{16\tan^2\theta + 16} d\theta = \int \frac{\sec^2\theta}{16\sec^2\theta} d\theta = \frac{1}{16}\int d\theta = \frac{\theta}{16} + c$$

$$= \frac{1}{16}\tan^{-1}\left(\frac{u}{4}\right) + c = \frac{1}{16}\tan^{-1}\left(\frac{2x^2 + 3}{4}\right) + c$$

Check the antiderivative: Apply the chain rule. Recall from first-semester calculus that $\dfrac{d}{dx}\tan^{-1}w = \dfrac{1}{1+w^2}\dfrac{dw}{dx}$ provided that $-\dfrac{\pi}{2} < \tan^{-1}w < \dfrac{\pi}{2}$.

$$\frac{d}{dx}\left[\frac{1}{16}\tan^{-1}\left(\frac{2x^2 + 3}{4}\right) + c\right] = \frac{1}{16}\frac{1}{1 + \left(\frac{2x^2 + 3}{4}\right)^2}\frac{d}{dx}\left(\frac{2x^2 + 3}{4}\right) = \frac{1}{16}\frac{1}{1 + \frac{4x^4 + 12x + 9}{16}}\frac{4x}{4}$$

$$= \frac{x}{16\left(1 + \frac{4x^4 + 12x + 9}{16}\right)} = \frac{1}{16 + 4x^4 + 12x + 9} = \frac{x}{4x^4 + 12x + 25}$$

⑱ $t^2 + 6t + 5 = (t + 3)^2 - 4$

Check: $(t + 3)^2 - 4 = t^2 + 2(t)(3) + 9 - 4 = t^2 + 6t + 5$

$$\int \frac{t}{t^2 + 6t + 5} dt = \int \frac{t}{(t + 3)^2 - 4} dt$$

$u = t + 3, \dfrac{du}{dt} = 1, du = dt$ Notes: $t = u - 3$; separate into two integrals.

$$\int \frac{t}{(t + 3)^2 - 4} dt = \int \frac{u - 3}{u^2 - 4} du = \int \frac{u}{u^2 - 4} du - 3\int \frac{du}{u^2 - 4}$$

Left integral: $w = u^2 - 4, \dfrac{dw}{du} = 2u, dw = 2udu, \dfrac{dw}{2} = udu$

Right integral: $u = 2\sec\theta, du = 2\sec\theta\tan\theta \, d\theta$

$$\int \frac{u}{u^2 - 4} du - 3\int \frac{du}{u^2 - 4} = \int \frac{dw}{2w} - 3\int \frac{2\sec\theta\tan\theta}{4\sec^2\theta - 4} d\theta$$

Answers with Full Solutions

$$= \frac{1}{2}\int \frac{dw}{w-4} - \frac{3(2)}{4}\int \frac{\sec\theta \tan\theta}{\tan^2\theta}d\theta = \frac{1}{2}\ln|w| - \frac{3}{2}\int \frac{\sec\theta}{\tan\theta}d\theta$$

$$= \frac{1}{2}\ln|u^2-4| - \frac{3}{2}\int \frac{1}{\cos\theta}\frac{\cos\theta}{\sin\theta}d\theta = \frac{1}{2}\ln|(t+3)^2-4| - \frac{3}{2}\int \csc\theta\, d\theta$$

$$= \frac{1}{2}\ln|(t+3)^2-4| - \frac{3}{2}\ln|\csc\theta - \cot\theta| + c$$

$$= \frac{1}{2}\ln|(t+3)^2-4| - \frac{3}{2}\ln\left|\csc\left(\sec^{-1}\frac{u}{2}\right) - \cot\left(\sec^{-1}\frac{u}{2}\right)\right| + c$$

$$= \frac{1}{2}\ln|(t+3)^2-4| - \frac{3}{2}\ln\left|\frac{u/2}{\sqrt{u^2/4-1}} - \frac{1}{\sqrt{u^2/4-1}}\right| + c$$

$$= \frac{1}{2}\ln|(t+3)^2-4| - \frac{3}{2}\ln\left|\frac{u/2-1}{\sqrt{u^2/4-1}}\right| + c = \frac{1}{2}\ln|(t+3)^2-4| - \frac{3}{2}\ln\left|\frac{u-2}{2\sqrt{u^2/4-1}}\right| + c$$

$$= \frac{1}{2}\ln|(t+3)^2-4| - \frac{3}{2}\ln\left|\frac{u-2}{\sqrt{u^2-4}}\right| + c = \frac{1}{2}\ln|(t+3)^2-4| - \frac{3}{2}\ln\left|\frac{u-2}{\sqrt{(u-2)(u+2)}}\right| + c$$

$$= \frac{1}{2}\ln|(t+3)^2-4| - \frac{3}{2}\ln\sqrt{\frac{u-2}{u+2}} + c = \frac{1}{2}\ln|(t+3)^2-4| - \frac{3}{4}\ln\left|\frac{u-2}{u+2}\right| + c$$

$$= \frac{1}{2}\ln|(t+3)^2-4| - \frac{3}{4}\ln\left|\frac{t+3-2}{t+3+2}\right| + c = \frac{1}{2}\ln|(t+3)^2-4| - \frac{3}{4}\ln\left|\frac{t+1}{t+5}\right| + c$$

$$= \frac{1}{2}\ln|(t+3)^2-4| - \frac{3}{4}\ln|t+1| + \frac{3}{4}\ln|t+5| + c$$

Notes: See the footnotes in Chapter 3 regarding inverse trig identities. We used $\ln\sqrt{p} = \ln p^{1/2} = \frac{1}{2}\ln p$.

Thought: Did you notice that $t^2 + 6t + 5$ factors as $(t+1)(t+5)$? Another way to solve this problem is to use the method of partial fractions (Chapter 6).

Check the answer: Apply the chain rule.

$$\frac{d}{dt}\left[\frac{1}{2}\ln[(t+3)^2-4] - \frac{3}{4}\ln(t+1) + \frac{3}{4}\ln(t+5) + c\right]$$

$$= \frac{1}{2[(t+3)^2-4]}\frac{d}{dt}[(t+3)^2-4] - \frac{3}{4(t+1)}\frac{d}{dt}(t+1) + \frac{3}{4(t+5)}\frac{d}{dt}(t+5)$$

$$= \frac{1}{2(t^2+6t+9-4)}\frac{d}{dt}(t^2+6t+9-4) - \frac{3}{4(t+1)} + \frac{3}{4(t+5)}$$

$$= \frac{2t+6}{2(t^2+6t+5)} - \frac{3}{4(t+1)} + \frac{3}{4(t+5)} = \frac{2t+6}{2(t+1)(t+5)} - \frac{3}{4(t+1)} + \frac{3}{4(t+5)}$$

$$= \frac{(2t+6)(2)}{4(t+1)(t+5)} - \frac{3(t+5)}{4(t+1)(t+5)} + \frac{3(t+1)}{4(t+5)(t+1)} = \frac{4t+12-(3t+15)+3t+3}{4(t+1)(t+5)}$$

$$= \frac{4t+12-3t-15+3t+3}{4(t^2+5t+t+5)} = \frac{4t}{4(t^2+6t+5)} = \frac{t}{t^2+6t+5}$$

⑲ $9y^2 - 30y + 50 = (3y-5)^2 + 25$

Check: $(3y-5)^2 + 25 = 9y^2 + 2(3y)(-5) + 25 + 25 = 9y^2 - 30y + 50$

$$\int_{y=5/3}^{10/3} \frac{dy}{\sqrt{9y^2-30y+50}} = \int_{y=5/3}^{10/3} \frac{dy}{\sqrt{(3y-5)^2+25}}$$

$u = 3y - 5, du = 3dy, \dfrac{du}{3} = dy$

At $y = 5/3$, $u = 3\left(\dfrac{5}{3}\right) - 5 = 5 - 5 = 0$. At $y = \dfrac{10}{3}$, $u = 3\left(\dfrac{10}{3}\right) - 5 = 10 - 5 = 5$.

$$\int_{y=5/3}^{10/3} \frac{dy}{\sqrt{(3y-5)^2+25}} = \frac{1}{3}\int_{u=0}^{5} \frac{du}{\sqrt{u^2+25}}$$

$u = 5\tan\theta, du = 5\sec^2\theta\, d\theta$

At $u = 0$, $\theta = \tan^{-1}\left(\dfrac{0}{5}\right) = 0$. At $u = 5$, $\theta = \tan^{-1}\left(\dfrac{5}{5}\right) = \tan^{-1} 1 = \dfrac{\pi}{4}$.

$$\frac{1}{3}\int_{u=0}^{5} \frac{du}{\sqrt{u^2+25}} = \frac{1}{3}\int_{\theta=0}^{\pi/4} \frac{5\sec^2\theta}{\sqrt{25\tan^2\theta+25}}d\theta = \frac{1}{3}\int_{\theta=0}^{\pi/4} \frac{5\sec^2\theta}{\sqrt{25\sec^2\theta}}d\theta$$

$$= \frac{1}{3}\int_{\theta=0}^{\pi/4} \frac{5\sec^2\theta}{5\sec\theta}d\theta = \frac{1}{3}\int_{\theta=0}^{\pi/4} \sec\theta\, d\theta = \frac{1}{3}[\ln|\sec\theta+\tan\theta|]_{\theta=0}^{\pi/4}$$

$$= \frac{1}{3}\ln\left|\sec\frac{\pi}{4}+\tan\frac{\pi}{4}\right| - \frac{1}{3}\ln|\sec 0+\tan 0| = \frac{1}{3}\ln|\sqrt{2}+1| - \frac{1}{3}\ln|1+0|$$

$$= \frac{1}{3}\ln(1+\sqrt{2}) - \frac{1}{3}\ln 1 = \frac{1}{3}\ln(1+\sqrt{2}) - 0 = \frac{\ln(1+\sqrt{2})}{3} \approx 0.294$$

Check the antiderivative: $f = \dfrac{1}{3}\ln(\sec\theta + \tan\theta) = \dfrac{1}{3}\ln\left[\sec\left(\tan^{-1}\dfrac{u}{5}\right) + \right.$

$\left.\tan\left(\tan^{-1}\dfrac{u}{5}\right)\right] = \dfrac{1}{3}\ln\left[\sqrt{1+\left(\dfrac{u}{5}\right)^2} + \dfrac{u}{5}\right] = \dfrac{1}{3}\ln\left[\sqrt{1+\left(\dfrac{3y-5}{5}\right)^2} + \dfrac{3y-5}{5}\right]$. See the footnotes

in Chapter 3 regarding the trig identities. Apply the chain rule.

$$\frac{d}{d\theta}\left\{\frac{1}{3}\ln\left[\sqrt{1+\left(\frac{3y-5}{5}\right)^2}+\frac{3y-5}{5}\right]\right\} = \frac{1}{3\left[\sqrt{1+\left(\frac{3y-5}{5}\right)^2}+\frac{3y-5}{5}\right]}\left[\frac{2\left(\frac{3y-5}{5}\right)\left(\frac{3}{5}\right)}{2\sqrt{1+\left(\frac{3y-5}{5}\right)^2}}+\frac{3}{5}\right]$$

$$= \frac{3/5}{3\left[\sqrt{1+\left(\frac{3y-5}{5}\right)^2}+\frac{3y-5}{5}\right]}\left[\frac{\left(\frac{3y-5}{5}\right)}{\sqrt{1+\left(\frac{3y-5}{5}\right)^2}}+1\right] = \frac{1/5}{\left[\sqrt{1+\left(\frac{3y-5}{5}\right)^2}+\frac{3y-5}{5}\right]}\left[\frac{\left(\frac{3y-5}{5}\right)}{\sqrt{1+\left(\frac{3y-5}{5}\right)^2}}+\frac{\sqrt{1+\left(\frac{3y-5}{5}\right)^2}}{\sqrt{1+\left(\frac{3y-5}{5}\right)^2}}\right]$$

$$= \frac{1}{5\left[\sqrt{1+\left(\frac{3y-5}{5}\right)^2}+\frac{3y-5}{5}\right]}\left[\frac{\sqrt{1+\left(\frac{3y-5}{5}\right)^2}+\left(\frac{3y-5}{5}\right)}{\sqrt{1+\left(\frac{3y-5}{5}\right)^2}}\right] = \frac{1}{5\sqrt{1+\left(\frac{3y-5}{5}\right)^2}} = \frac{1}{\sqrt{25+9y^2-30y+25}} = \frac{1}{\sqrt{9y^2-30y+50}}$$

⑳ $\cos^2\theta + 2\sin\theta + 2 = 1 - \sin^2\theta + 2\sin\theta + 2 = 3 + 2\sin\theta - \sin^2\theta$

Note: $\cos^2\theta = 1 - \sin^2\theta$ because $\sin^2\theta + \cos^2\theta = 1$.

$3 + 2\sin\theta - \sin^2\theta = 4 - (1 - \sin\theta)^2 = 4 - (\sin\theta - 1)^2$

Note: $(1 - \sin\theta)^2 = (\sin\theta - 1)^2$ for the same reason that $(-p)^2 = p^2$.

Check: $4 - (\sin\theta - 1)^2 = 4 - (\sin^2\theta - 2\sin\theta + 1) = 4 - \sin^2\theta + 2\sin\theta - 1$
$= 3 + 2\sin\theta - \sin^2\theta = 3 + 2\sin\theta - (1 - \cos^2\theta) = 3 + 2\sin\theta - 1 + \cos^2\theta$
$= 2 + 2\sin\theta + \cos^2\theta$

$$\int_{\theta=0}^{\pi/2} \frac{\cos\theta}{\cos^2\theta + 2\sin\theta + 2}\,d\theta = \int_{\theta=0}^{\pi/2} \frac{\cos\theta}{4 - (\sin\theta - 1)^2}\,d\theta$$

$u = \sin\theta - 1$, $du = \cos\theta\,d\theta$

At $\theta = 0$, $u = \sin 0 - 1 = 0 - 1 = -1$. At $\theta = \frac{\pi}{2}$, $u = \sin\frac{\pi}{2} - 1 = 1 - 1 = 0$.

$$\int_{\theta=0}^{\pi/2} \frac{\cos\theta}{4 - (\sin\theta-1)^2}\,d\theta = \int_{u=-1}^{0} \frac{du}{4 - u^2}$$

$u = 2\sin\varphi$, $du = 2\cos\varphi\,d\varphi$

At $u = -1$, $\varphi = \sin^{-1}\left(-\frac{1}{2}\right) = -\frac{\pi}{6}$. At $u = 0$, $\varphi = \sin^{-1}\left(\frac{0}{2}\right) = 0$.

$$\int_{u=-1}^{0} \frac{du}{4-u^2} = \int_{\varphi=-\pi/6}^{0} \frac{2\cos\varphi}{4 - 4\sin^2\varphi}\,d\varphi = \int_{\varphi=-\pi/6}^{0} \frac{2\cos\varphi}{4\cos^2\varphi}\,d\varphi$$

$$= \frac{1}{2}\int_{\varphi=-\pi/6}^{0} \frac{d\varphi}{\cos\varphi} = \frac{1}{2}\int_{\varphi=-\pi/6}^{0} \sec\varphi\,d\varphi = \frac{1}{2}[\ln|\sec\varphi + \tan\varphi|]_{\varphi=-\pi/6}^{0}$$

$$= \frac{1}{2}\ln|\sec 0 + \tan 0| - \frac{1}{2}\ln\left|\sec\left(-\frac{\pi}{6}\right) + \tan\left(-\frac{\pi}{6}\right)\right| = \frac{1}{2}\ln|1+0| - \frac{1}{2}\ln\left|\frac{2}{\sqrt{3}} - \frac{1}{\sqrt{3}}\right|$$

$$= \frac{1}{2}\ln 1 - \frac{1}{2}\ln\left(\frac{1}{\sqrt{3}}\right) = 0 + \frac{1}{2}\ln\sqrt{3} = \frac{1}{4}\ln 3 \approx 0.275$$

Note: We applied the rules $\ln\left(\frac{1}{p}\right) = \ln p^{-1} = -\ln p$ and $\ln\sqrt{q} = \ln q^{1/2} = \frac{1}{2}\ln q$.

Check the antiderivative: But first, we'll do some algebra with the antiderivative.

$$\frac{1}{2}\ln(\sec\varphi + \tan\varphi) = \frac{1}{2}\ln\left[\sec\left(\sin^{-1}\frac{u}{2}\right) + \tan\left(\sin^{-1}\frac{u}{2}\right)\right]$$

$$= \frac{1}{2}\ln\left[\frac{1}{\sqrt{1-\left(\frac{u}{2}\right)^2}} + \frac{\frac{u}{2}}{\sqrt{1-\left(\frac{u}{2}\right)^2}}\right] = \frac{1}{2}\ln\left(\frac{2}{2\sqrt{1-\frac{u^2}{4}}} + \frac{u}{2\sqrt{1-\frac{u^2}{4}}}\right) = \frac{1}{2}\ln\left(\frac{2+u}{\sqrt{4-u^2}}\right) = \frac{1}{2}\ln\left[\frac{2+u}{\sqrt{(2+u)(2-u)}}\right]$$

$$= \frac{1}{2}\ln\sqrt{\frac{2+u}{2-u}} = \frac{1}{4}\ln\left|\frac{2+u}{2-u}\right| = \frac{1}{4}\ln|2+u| - \frac{1}{4}\ln|2-u|$$

(We used the rule $\ln\sqrt{q} = \ln q^{1/2} = \frac{1}{2}\ln q$.)

$$= \frac{1}{4}\ln|2+\sin\theta - 1| - \frac{1}{4}\ln|2 - (\sin\theta - 1)| = \frac{1}{4}\ln(1+\sin\theta) - \frac{1}{4}\ln|2-\sin\theta + 1|$$

$$= \frac{1}{4}\ln(1+\sin\theta) - \frac{1}{4}\ln(3-\sin\theta)$$

Notes: This is just the **antiderivative**. See the footnotes in Chapter 3 regarding the trig identities. Apply the chain rule.

$$\frac{d}{d\theta}\left[\frac{1}{4}\ln(1+\sin\theta) - \frac{1}{4}\ln(3-\sin\theta)\right]$$

$$= \frac{1}{4(1+\sin\theta)}\frac{d}{d\theta}(1+\sin\theta) - \frac{1}{4(3-\sin\theta)}\frac{d}{d\theta}(3-\sin\theta)$$

$$= \frac{\cos\theta}{4(1+\sin\theta)} - \frac{-\cos\theta}{4(3-\sin\theta)} = \frac{\cos\theta}{4(1+\sin\theta)} + \frac{\cos\theta}{4(3-\sin\theta)}$$

$$= \frac{\cos\theta(3-\sin\theta)}{4(1+\sin\theta)(3-\sin\theta)} + \frac{\cos\theta(1+\sin\theta)}{4(3-\sin\theta)(1+\sin\theta)}$$

$$= \frac{3\cos\theta - \sin\theta\cos\theta}{4(3-\sin\theta+3\sin\theta-\sin^2\theta)} + \frac{\cos\theta + \sin\theta\cos\theta}{4(3-\sin\theta+3\sin\theta-\sin^2\theta)}$$

$$= \frac{4\cos\theta}{4(3+2\sin\theta-\sin^2\theta)} = \frac{\cos\theta}{3+2\sin\theta-\sin^2\theta} = \frac{\cos\theta}{3+2\sin\theta-(1-\cos^2\theta)}$$

$$= \frac{\cos\theta}{3+2\sin\theta-1+\cos^2\theta} = \frac{\cos\theta}{2+2\sin\theta+\cos^2\theta}$$

Answers with Full Solutions

Note: Some of the problems in this chapter would be simpler if applying the method of **partial fractions** (Chapter 6) instead of completing the square. This is often the case when you're able to factor the denominator. In this problem, the denominator can be factored as

$$\cos^2\theta + 2\sin\theta + 2 = 1 - \sin^2\theta + 2\sin\theta + 2 = 3 + 2\sin\theta - \sin^2\theta$$
$$= (3 - \sin\theta)(1 + \sin\theta)$$

Chapter 6 Solutions

❶ $\dfrac{1}{4t^2-25} = \dfrac{a}{2t-5} + \dfrac{b}{2t+5} \to 1 = a(2t+5) + b(2t-5)$

$1 = 2at + 5a + 2bt - 5b = 2(a+b)t + 5a - 5b$

$0 = 2(a+b)t \quad , \quad 1 = 5a - 5b \quad \to \quad a + b = 0 \to b = -a$

$1 = 5a - 5(-a) = 5a + 5a = 10a \to \dfrac{1}{10} = a \quad \to \quad b = -a = -\dfrac{1}{10}$

Check: $\dfrac{1/10}{2t-5} + \dfrac{-1/10}{2t+5} = \dfrac{\tfrac{1}{10}(2t+5) - \tfrac{1}{10}(2t-5)}{(2t-5)(2t+5)} = \dfrac{\tfrac{t}{5} + \tfrac{1}{2} - \tfrac{t}{5} + \tfrac{1}{2}}{4t^2 - 25} = \dfrac{1}{4t^2 - 25}$

$\displaystyle\int \dfrac{dt}{4t^2 - 25} = \dfrac{1}{10}\int \dfrac{dt}{2t-5} - \dfrac{1}{10}\int \dfrac{dt}{2t+5}$

$u = 2t - 5, du = 2dt, \dfrac{du}{2} = dt, w = 2t + 5, dw = 2dt, \dfrac{dw}{2} = dt$

$\dfrac{1}{10}\displaystyle\int \dfrac{dt}{2t-5} - \dfrac{1}{10}\int \dfrac{dt}{2t+5} = \dfrac{1}{10}\int \dfrac{du}{2u} - \dfrac{1}{10}\int \dfrac{dw}{2w} = \dfrac{1}{20}\ln|u| - \dfrac{1}{20}\ln|w| + c$

$= \dfrac{1}{20}\ln|2t-5| - \dfrac{1}{20}\ln|2t+5| + c = \dfrac{1}{20}\ln\left|\dfrac{2t-5}{2t+5}\right| + c$

Alternate answer: $\dfrac{1}{10}\ln\sqrt{\dfrac{2t-5}{2t+5}} + c$

Check the answer: Take a derivative with respect to t. Apply the chain rule.

$\dfrac{d}{dt}\left[\dfrac{1}{20}\ln(2t-5) - \dfrac{1}{20}\ln(2t+5) + c\right]$

$= \dfrac{1}{20(2t-5)}\dfrac{d}{dt}(2t-5) - \dfrac{1}{20(2t+5)}\dfrac{d}{dt}(2t+5) = \dfrac{1}{10(2t-5)} - \dfrac{1}{10(2t+5)}$

$= \dfrac{2t+5}{10(2t-5)(2t+5)} - \dfrac{2t-5}{10(2t+5)(2t-5)} = \dfrac{2t+5 - (2t-5)}{10(4t^2-25)}$

$= \dfrac{2t+5-2t+5}{10(4t^2-25)} = \dfrac{10}{10(4t^2-25)} = \dfrac{1}{4t^2-25}$

❷ $\dfrac{1}{6y^2 - 7y - 20} = \dfrac{a}{2y-5} + \dfrac{b}{3y+4} \to 1 = a(3y+4) + b(2y-5)$

$1 = 3ay + 4a + 2by - 5b = (3a + 2b)y + 4a - 5b$

$0 = (3a+2b)y \quad , \quad 1 = 4a - 5b \quad \to \quad 3a + 2b = 0 \to 2b = -3a \to b = -\dfrac{3}{2}a$

$1 = 4a - 5\left(-\dfrac{3}{2}a\right) = 4a + \dfrac{15a}{2} = \dfrac{8a}{2} + \dfrac{15a}{2} = \dfrac{23a}{2} \to \dfrac{2}{23} = a$

Answers with Full Solutions

$b = -\dfrac{3}{2}a = -\dfrac{3}{2}\dfrac{2}{23}a = -\dfrac{3}{23}$

Check: $\dfrac{2/23}{2y-5} + \dfrac{-3/23}{3y+4} = \dfrac{\frac{2}{23}(3y+4) - \frac{3}{23}(2y-5)}{(2y-5)(3y+4)} = \dfrac{\frac{6y}{23} + \frac{8}{23} - \frac{6y}{23} + \frac{15}{23}}{6y^2 - 7y - 20} = \dfrac{23/23}{6y^2 - 7y - 20} = \dfrac{1}{6y^2 - 7y - 20}$

$\displaystyle\int \dfrac{dy}{6y^2 - 7y - 20} = \dfrac{2}{23}\int \dfrac{dy}{2y - 5} - \dfrac{3}{23}\int \dfrac{dy}{3y + 4}$

$u = 2y - 5, du = 2dy, \dfrac{du}{2} = dy, w = 3y + 4, dw = 3dy, \dfrac{dw}{3} = dy$

$\dfrac{2}{23}\int \dfrac{dy}{2u} - \dfrac{3}{23}\int \dfrac{dy}{3w} = \dfrac{1}{23}\int \dfrac{du}{u} - \dfrac{1}{23}\int \dfrac{dw}{w} = \dfrac{1}{23}\ln|u| - \dfrac{1}{23}\ln|w| + c$

$= \dfrac{1}{23}\ln|2y - 5| - \dfrac{1}{23}\ln|3y + 4| + c = \dfrac{1}{23}\ln\left|\dfrac{2y-5}{3y+4}\right| + c$

Note: $\ln\left|\dfrac{12y-30}{12y+16}\right| = \ln\left|\dfrac{6(2y-5)}{4(3y+4)}\right| = \ln\left(\dfrac{6}{4}\right) + \ln\left|\dfrac{2y-5}{3y+4}\right| = \ln\left|\dfrac{2y-5}{3y+4}\right| +$ a constant. The answer can thus be expressed in a multitude of equivalent, but different-looking forms. Check the answer: Take a derivative with respect to y. Apply the chain rule.

$\dfrac{d}{dy}\left[\dfrac{1}{23}\ln(2y-5) - \dfrac{1}{23}\ln(3y+4) + c\right] = \dfrac{1}{23(2y-5)}\dfrac{d}{dy}(2y-5) - \dfrac{1}{23(3y+4)}\dfrac{d}{dy}(3y+4)$

$= \dfrac{2}{23(2y-5)} - \dfrac{3}{23(3y+4)} = \dfrac{2(3y+4)}{23(2y-5)(3y+4)} - \dfrac{3(2y-5)}{10(3y+4)(2y-5)}$

$= \dfrac{6y + 8 - (6y - 15)}{10(6y^2 - 7y - 20)} = \dfrac{6y + 8 - 6y + 15}{23(6y^2 - 7y - 20)} = \dfrac{23}{23(6y^2 - 7y - 20)} = \dfrac{1}{6y^2 - 7y - 20}$

❸ $\dfrac{14x + 3}{4x^2 + 21x - 18} = \dfrac{a}{x + 6} + \dfrac{b}{4x - 3} \to 14x + 3 = a(4x - 3) + b(x + 6)$

$14x + 3 = 4ax - 3a + bx + 6b = (4a + b)x - 3a + 6b$

$14 = 4a + b$, $3 = -3a + 6b \to 1 = -a + 2b \to 1 + a = 2b \to \dfrac{1+a}{2} = b$

$14 = 4a + \dfrac{1+a}{2} \to 28 = 8a + 1 + a \to 28 - 1 = 9a \to 27 = 9a \to \dfrac{27}{9} = 3 = a$

$b = \dfrac{1+3}{2} = \dfrac{4}{2} = 2$

Check: $\dfrac{3}{x+6} + \dfrac{2}{4x-3} = \dfrac{3(4x-3)}{(x+6)(4x-3)} + \dfrac{2(x+6)}{(x+6)(4x-3)} = \dfrac{12x - 9 + 2x + 12}{4x^2 + 24x - 3x - 18} = \dfrac{14x+3}{4x^2 + 21x - 18}$

$\displaystyle\int \dfrac{14x + 3}{4x^2 + 21x - 18}dx = 3\int \dfrac{dx}{x + 6} + 2\int \dfrac{dx}{4x - 3}$

$u = x + 6, du = dx, w = 4x - 3, dw = 4dy, \dfrac{dw}{4} = dy$

$$3\int \frac{du}{u} + 2\int \frac{dw}{4w} = 3\int \frac{du}{u} + \frac{1}{2}\int \frac{dw}{w} = 3\ln|u| + \frac{1}{2}\ln|w| + c$$

$$= 3\ln|x+6| + \frac{1}{2}\ln|4x-3| + c$$

Check the answer: Take a derivative with respect to x. Apply the chain rule.

$$\frac{d}{dx}\left[3\ln(x+6) + \frac{1}{2}\ln(4x-3) + c\right] = \frac{3}{x+6}\frac{d}{dx}(x+6) + \frac{1}{2(4x-3)}\frac{d}{dx}(4x-3)$$

$$= \frac{3(1)}{x+6} + \frac{4}{2(4x-3)} = \frac{3}{x+6} + \frac{2}{4x-3} = \frac{3(4x-3)}{(x+6)(4x-3)} + \frac{2(x+6)}{(4x-3)(x+6)}$$

$$= \frac{12x-9+2x+12}{4x^2+24x-3x-18} = \frac{14x+3}{4x^2+21x-18}$$

④ $\dfrac{7z^2+29z-36}{z^3+z^2-6z} = \dfrac{a}{z} + \dfrac{b}{z-2} + \dfrac{c}{z+3}$

$7z^2 + 29z - 36 = a(z-2)(z+3) + bz(z+3) + cz(z-2)$

$7z^2 + 29z - 36 = a(z^2+z-6) + bz^2 + 3bz + cz^2 - 2cz$

$7z^2 + 29z - 36 = az^2 + az - 6a + bz^2 + 3bz + cz^2 - 2cz$

$7z^2 + 29z - 36 = (a+b+c)z^2 + (a+3b-2c)z - 6a$

$7 = a+b+c$, $29 = a+3b-2c$, $-36 = -6a$

$\dfrac{36}{6} = 6 = a$ → $7 = 6+b+c$, $29 = 6+3b-2c$

$1 = b+c$, $23 = 3b-2c$ → $b = 1-c$ → $23 = 3(1-c) - 2c$

$23 = 3 - 3c - 2c = 3 - 5c$ → $20 = -5c$ → $-\dfrac{20}{5} = -4 = c$

$b = 1 - c = 1 - (-4) = 1 + 4 = 5$

Check: $\dfrac{6}{z} + \dfrac{5}{z-2} + \dfrac{-4}{z+3} = \dfrac{6(z^2+z-6)+5z(z+3)-4z(z-2)}{z(z^2+z-6)} = \dfrac{6z^2+6z-36+5z^2+15z-4z^2+8z}{z^3+z^2-6z} = \dfrac{7z^2+29z-36}{z^3+z^2-6z}$

$$\int \frac{7z^2+29z-36}{z^3+z^2-6z}dz = 6\int \frac{dz}{z} + 5\int \frac{dz}{z-2} - 4\int \frac{dz}{z+3}$$

$$= 6\ln|z| + 5\ln|z-2| - 4\ln|z+3| + c$$

Note: If you let $u = z - 2$, for example, $du = dz$. These integrals are a bit simpler because the du's don't involve any coefficients (or other factors).

Check the answer: Take a derivative with respect to z. Apply the chain rule.

$$\frac{d}{dz}[6\ln z + 5\ln(z-2) - 4\ln(z+3) + c] = \frac{6}{z} + \frac{5}{z-2}\frac{d}{dz}(z-2) - \frac{4}{z+3}\frac{d}{dz}(z+3)$$

Answers with Full Solutions

$$= \frac{6}{z} + \frac{5}{z-2} - \frac{4}{z+3} = \frac{6(z^2+z-6)+5z(z+3)-4z(z-2)}{z(z^2+z-6)}$$

$$= \frac{6z^2+6z-36+5z^2+15z-4z^2+8z}{z^3+z^2-6z} = \frac{7z^2+29z-36}{z^3+z^2-6z}$$

❺ $u = e^w, du = e^w dw$

$$\int \frac{e^w dw}{e^{2w}-9} = \int \frac{du}{u^2-9}$$

$$\frac{1}{u^2-9} = \frac{a}{u-3} + \frac{b}{u+3}$$

$$1 = a(u+3) + b(u-3) = au + 3a + bu - 3b = (a+b)u + 3a - 3b$$

$$0 = (a+b)u \quad , \quad 1 = 3a - 3b \quad \to \quad 0 = a+b \to b = -a$$

$$1 = 3a - 3(-a) = 3a + 3a = 6a \to \frac{1}{6} = a \quad \to \quad b = -a = -\frac{1}{6}$$

Check: $\frac{1/6}{u-3} + \frac{-1/6}{u+3} = \frac{\frac{1}{6}(u+3)-\frac{1}{6}(u-3)}{(u-3)(u+3)} = \frac{\frac{u}{6}+\frac{3}{6}-\frac{u}{6}+\frac{3}{6}}{u^2-9} = \frac{6/6}{u^2-9} = \frac{1}{u^2-9}$

$$\int \frac{du}{u^2-9} = \frac{1}{6}\int \frac{du}{u-3} - \frac{1}{6}\int \frac{du}{u+3} = \frac{1}{6}\ln|u-3| - \frac{1}{6}\ln|u+3| + c$$

$$= \frac{1}{6}\ln|e^w-3| - \frac{1}{6}\ln|e^w+3| + c = \frac{1}{6}\ln\left|\frac{e^w-3}{e^w+3}\right| + c$$

Note: If you let $t = u - 3$, for example, $dt = du$. These integrals are a bit simpler because the dt's don't involve any coefficients (or other factors).

Check the answer: Take a derivative with respect to w. Apply the chain rule.

$$\frac{d}{dw}\left[\frac{1}{6}\ln(e^w-3) - \frac{1}{6}\ln(e^w+3) + c\right] = \frac{1}{6(e^w-3)}\frac{d}{dw}(e^w-3) - \frac{1}{6(e^w+3)}\frac{d}{dw}(e^w+3)$$

$$= \frac{e^w}{6(e^w-3)} - \frac{e^w}{6(e^w+3)} = \frac{e^w(e^w+3)}{6(e^w-3)(e^w+3)} - \frac{e^w(e^w-3)}{6(e^w-3)(e^w+3)}$$

$$= \frac{e^{2w}+3e^w-(e^{2w}-3e^w)}{6(e^{2w}+3e^w-3e^w-9)} = \frac{e^{2w}+3e^w-e^{2w}+3e^w}{6(e^{2w}-9)} = \frac{6e^w}{6(e^{2w}-9)} = \frac{e^w}{e^{2w}-9}$$

❻ Note: This is similar to Example 3 since the numerator has a higher degree than the denominator. The first step is to divide $x^3 - 5x^2 + 11x - 12$ by $x - 2$. You can either do polynomial long division or use the strategy from Example 3. The result of this division will have the form $x^2 + ax + b + \frac{c}{x-2}$. Write out the division problem.

$$\frac{x^3-5x^2+11x-12}{x-2} = x^2 + ax + b + \frac{c}{x-2}$$

190

Multiply by $x - 2$ on both sides to rewrite it as a multiplication problem.

$x^3 - 5x^2 + 11x - 12 = (x-2)\left(x^2 + ax + b + \dfrac{c}{x-2}\right)$

$x^3 - 5x^2 + 11x - 12 = x^3 - 2x^2 + ax^2 - 2ax + bx - 2b + c$

Equate coefficients of the same power on both sides.

$-5 = -2 + a$, $\quad 11 = -2a + b$, $\quad -12 = -2b + c$

$-5 + 2 = a = -3 \quad \rightarrow \quad 11 = -2(-3) + b \rightarrow 11 = 6 + b \rightarrow 11 - 6 = 5 = b$

$-12 = -2(5) + c \rightarrow -12 = -10 + c \rightarrow -12 + 10 = -2 = c$

$\dfrac{x^3 - 5x^2 + 11x - 12}{x - 2} = x^2 + ax + b + \dfrac{c}{x-2} = x^2 - 3x + 5 - \dfrac{2}{x-2}$

$\displaystyle\int \dfrac{x^3 - 5x^2 + 11x - 12}{x - 2}\,dx = \int x^2\,dx - 3\int x\,dx + \int 5\,dx - 2\int \dfrac{dx}{x-2}$

$= \dfrac{x^3}{3} - \dfrac{3x^2}{2} + 5x - 2\ln|x-2| + c$

Check the answer: Take a derivative with respect to x. Apply the chain rule.

$\dfrac{d}{dx}\left[\dfrac{x^3}{3} - \dfrac{3x^2}{2} + 5x - 2\ln|x-2| + c\right] = x^2 - 3x + 5 - \dfrac{2}{x-2}\dfrac{d}{dx}(x-2)$

$= x^2 - 3x + 5 - \dfrac{2(1)}{x-2} = \dfrac{x^2(x-2) - 3x(x-2) + 5(x-2) - 2}{x-2}$

$= \dfrac{x^3 - 2x^2 - 3x^2 + 6x + 5x - 10 - 2}{x - 2} = \dfrac{x^3 - 5x^2 + 11x - 12}{x - 2}$

❼ $u = \cos\theta$, $du = -\sin\theta\,d\theta$

At $\theta = \dfrac{\pi}{6}$, $u = \cos\dfrac{\pi}{6} = \dfrac{\sqrt{3}}{2}$. At $\theta = \dfrac{\pi}{3}$, $u = \cos\dfrac{\pi}{3} = \dfrac{1}{2}$.

$\displaystyle\int_{\theta = \pi/6}^{\pi/3} \dfrac{\sin\theta}{4\cos\theta - \cos^2\theta}\,d\theta = -\int_{u=\sqrt{3}/2}^{1/2} \dfrac{du}{4u - u^2}$

Note: The factoring is a bit different here than in other problems: $4u - u^2 = u(4-u)$. The factors are u and $4 - u$.

$\dfrac{1}{4u - u^2} = \dfrac{a}{u} + \dfrac{b}{4-u} \quad \rightarrow \quad 1 = a(4-u) + bu = 4a - au + bu = (b-a)u + 4a$

$0 = (b-a)u$, $\quad 1 = 4a \quad \rightarrow \quad \dfrac{1}{4} = a \quad \rightarrow \quad b = a = \dfrac{1}{4}$

Check: $\dfrac{1/4}{u} + \dfrac{1/4}{4-u} = \dfrac{\tfrac{1}{4}(4-u) + \tfrac{1}{4}u}{u(4-u)} = \dfrac{1 - \tfrac{1}{4}u + \tfrac{1}{4}u}{4u - u^2} = \dfrac{1}{4u - u^2}$

Answers with Full Solutions

$$-\int_{u=\sqrt{3}/2}^{1/2} \frac{du}{4u - u^2} = -\frac{1}{4}\int_{u=\sqrt{3}/2}^{1/2} \frac{du}{u} - \frac{1}{4}\int_{u=\sqrt{3}/2}^{1/2} \frac{du}{4 - u}$$

$w = 4 - u, dw = -du, -dw = du$

At $u = \frac{\sqrt{3}}{2}, w = 4 - \frac{\sqrt{3}}{2}$. At $u = \frac{1}{2}, w = 4 - \frac{1}{2} = \frac{8}{2} - \frac{1}{2} = \frac{7}{2}$.

$$-\frac{1}{4}\int_{u=\sqrt{3}/2}^{1/2} \frac{du}{u} + \frac{1}{4}\int_{u=\sqrt{3}/2}^{1/2} \frac{du}{4 - u} = -\frac{1}{4}[\ln|u|]_{u=1/2}^{\sqrt{3}/2} + \frac{1}{4}\int_{u=4-\sqrt{3}/2}^{7/2} \frac{dw}{w}$$

$$= -\frac{1}{4}\ln\left(\frac{1}{2}\right) + \frac{1}{4}\ln\left(\frac{\sqrt{3}}{2}\right) + \frac{1}{4}[\ln|w|]_{w=4-\sqrt{3}/2}^{7/2} = \frac{1}{4}\ln\left(\frac{\sqrt{3}}{2} \div \frac{1}{2}\right) + \frac{1}{4}\ln\left(\frac{7}{2}\right) - \frac{1}{4}\ln\left(4 - \frac{\sqrt{3}}{2}\right)$$

$$= \frac{1}{4}\ln\left(\frac{\sqrt{3}}{2} \times \frac{2}{1}\right) + \frac{1}{4}\ln\left(\frac{7}{2}\right) - \frac{1}{4}\ln\left(\frac{8}{2} - \frac{\sqrt{3}}{2}\right) = \frac{1}{4}\ln\sqrt{3} + \frac{1}{4}\ln\left(\frac{7}{2}\right) - \frac{1}{4}\ln\left(\frac{8 - \sqrt{3}}{2}\right)$$

$$= \frac{1}{4}\ln\sqrt{3} + \frac{1}{4}\ln\left(\frac{7}{2} \div \frac{8 - \sqrt{3}}{2}\right) = \frac{1}{4}\ln\sqrt{3} + \frac{1}{4}\ln\left(\frac{7}{2} \times \frac{2}{8 - \sqrt{3}}\right)$$

$$= \frac{1}{4}\ln\sqrt{3} + \frac{1}{4}\ln\left(\frac{7}{8 - \sqrt{3}}\right) = \frac{1}{4}\ln\left(\frac{7\sqrt{3}}{8 - \sqrt{3}}\right) \approx 0.165$$

Alternate answers: $\frac{1}{4}\ln\left(\frac{21}{8\sqrt{3}-3}\right) = \frac{1}{4}\ln\left(\frac{56\sqrt{3}+21}{61}\right)$

Note: We used the rule $\ln\left(\frac{p}{q}\right) = \ln p - \ln q$.

Check the antiderivative: The antiderivative is $-\frac{1}{4}\ln u + \frac{1}{4}\ln(4 - u) = -\frac{1}{4}\ln(\cos\theta) + \frac{1}{4}\ln(4 - \cos\theta)$. Take a derivative with respect to θ. Apply the chain rule.

$$\frac{d}{d\theta}\left[-\frac{1}{4}\ln(\cos\theta) + \frac{1}{4}\ln(4 - \cos\theta)\right] = -\frac{1}{4\cos\theta}\frac{d}{d\theta}\cos\theta + \frac{1}{4(4 - \cos\theta)}\frac{d}{d\theta}(4 - \cos\theta)$$

$$= \frac{\sin\theta}{4\cos\theta} + \frac{\sin\theta}{4(4 - \cos\theta)} = \frac{\sin\theta(4 - \cos\theta)}{4\cos\theta(4 - \cos\theta)} + \frac{\sin\theta\cos\theta}{4(4 - \cos\theta)\cos\theta}$$

$$= \frac{4\sin\theta - \sin\theta\cos\theta + \sin\theta\cos\theta}{16\cos\theta - 4\cos^2\theta} = \frac{4\sin\theta}{16\cos\theta - 4\cos^2\theta} = \frac{\sin\theta}{4\cos\theta - \cos^2\theta}$$

❽ Note: This is similar to Example 3 since the numerator has a higher degree than the denominator. The first step is to divide x^2 by $x + 3$. You can either do polynomial long division or use the strategy from Example 3. The result of this division will have the form $x + a + \frac{b}{x+3}$. Write out the division problem.

$$\frac{x^2}{x+3} = x + a + \frac{b}{x+3}$$

Multiply by $x + 3$ on both sides to rewrite it as a multiplication problem.

$$x^2 = (x+3)\left(x + a + \frac{b}{x+3}\right)$$

$$x^2 = x^2 + 3x + ax + 3a + b = x^2 + (a+3)x + 3a + b$$

Equate coefficients of the same power on both sides.

$$0 = a + 3, \quad 0 = 3a + b$$

$$a = -3 \quad \to \quad 0 = 3(-3) + b \to 0 = -9 + b \to 9 = b$$

$$\frac{x^2}{x+3} = x + a + \frac{b}{x+3} = x - 3 + \frac{9}{x+3}$$

$$\int_{x=0}^{3} \frac{x^2}{x+3}\,dx = \int_{x=0}^{3}\left(x - 3 + \frac{9}{x+3}\right)dx = \int_{x=0}^{3} x\,dx - \int_{x=0}^{3} 3\,dx + \int_{x=0}^{3} \frac{9}{x+3}\,dx$$

$$= \left[\frac{x^2}{2} - 3x + 9\ln|x+3|\right]_{x=0}^{3} = \frac{3^2}{2} - 3(3) + 9\ln(3+3) - \frac{0^2}{2} + 3(0) - 9\ln(0+3)$$

$$= \frac{9}{2} - 9 + 9\ln 6 - 0 + 0 - 9\ln 3 = \frac{9}{2} - \frac{18}{2} + 9\ln\left(\frac{6}{3}\right) = -\frac{9}{2} + 9\ln 2 \approx 1.74$$

Note: $9\ln 6 - 9\ln 3 = 9\ln\left(\frac{6}{3}\right)$ according to $\ln\left(\frac{p}{q}\right) = \ln p - \ln q$.

Check the antiderivative: The antiderivative is $\frac{x^2}{2} - 3x + 9\ln(x+3)$. Take a derivative with respect to x. Apply the chain rule.

$$\frac{d}{dx}\left[\frac{x^2}{2} - 3x + 9\ln(x+3)\right] = x - 3 + \frac{9}{x+3}\frac{d}{dx}(x+3) = x - 3 + \frac{9(1)}{x+3}$$

$$= \frac{x(x+3)}{x+3} - \frac{3(x+3)}{x+3} + \frac{9}{x+3} = \frac{x^2 + 3x - (3x+9) + 9}{x+3} = \frac{x^2 + 3x - 3x - 9 + 9}{x+3}$$

$$= \frac{x^2}{x+3}$$

Chapter 7 Solutions

❶ $u = \ln z, dv = dz \rightarrow du = \dfrac{dz}{z}, v = z$

$\displaystyle\int \ln z \, dz = (\ln z)(z) - \int z \dfrac{dz}{z} = z \ln z - \int dz = z \ln z - z + c$

Check the answer: Apply the product rule.

$\dfrac{d}{dz}(z \ln z - z + c) = z\dfrac{d}{dz}\ln z + \ln z \dfrac{d}{dz} z - 1 = z\left(\dfrac{1}{z}\right) + (\ln z)(1) - 1$

$= 1 + \ln z - 1 = \ln z$

❷ $u = \tan^{-1} y, dv = dy \rightarrow du = \dfrac{dy}{1+y^2}, v = y$

$\displaystyle\int \tan^{-1} y \, dy = (\tan^{-1} y)(y) - \int y \dfrac{dy}{1+y^2} = y \tan^{-1} y - \int \dfrac{y}{1+y^2} dy$

$w = 1 + y^2, dw = 2y dy, \dfrac{dw}{2} = y dy$

$y \tan^{-1} y - \displaystyle\int \dfrac{y}{1+y^2} dy = y \tan^{-1} y - \dfrac{1}{2}\int \dfrac{dw}{w} = y \tan^{-1} y - \dfrac{\ln|w|}{2} + c$

$= y \tan^{-1} y - \dfrac{\ln|1+y^2|}{2} + c$

Check the answer: Apply the chain rule and the product rule.

$\dfrac{d}{dy}\left[y \tan^{-1} y - \dfrac{\ln(1+y^2)}{2} + c\right] = y\dfrac{d}{dy}\tan^{-1} y + \tan^{-1} y \dfrac{d}{dy} y - \dfrac{1}{2(1+y^2)}\dfrac{d}{dy}(1+y^2)$

$= y\left(\dfrac{1}{1+y^2}\right) + (\tan^{-1} y)(1) - \dfrac{1}{2(1+y^2)}(2y) = \dfrac{y}{1+y^2} + \tan^{-1} y - \dfrac{y}{1+y^2} = \tan^{-1} y$

❸ $u = \sec^{-1} x, dv = dx \rightarrow du = \dfrac{dx}{x\sqrt{x^2-1}}, v = x$

Note: From the limits of integration, $\sec^{-1} x$ lies in Quadrant I.

$\displaystyle\int_{x=1}^{2} \sec^{-1} x \, dx = [(\sec^{-1} x)(x)]_{x=1}^{2} - \int_{x=1}^{2} x \dfrac{dx}{x\sqrt{x^2-1}} = [x\sec^{-1} x]_{x=1}^{2} - \int_{x=1}^{2} \dfrac{dx}{\sqrt{x^2-1}}$

Note: With a **definite integral**, remember to evaluate uv over the integration limits.

$x = \sec\theta, dx = \sec\theta\tan\theta \, d\theta$ (recall Chapter 3)

At $x = 1, \theta = \sec^{-1} 1 = 0$. At $x = 2, \theta = \sec^{-1} 2 = \dfrac{\pi}{3}$.

194

$$[x\sec^{-1}x]_{x=1}^2 - \int_{x=1}^{2}\frac{dx}{\sqrt{x^2-1}} = 2\sec^{-1}2 - 1\sec^{-1}1 - \int_{\theta=0}^{\pi/3}\frac{\sec\theta\tan\theta}{\sqrt{\sec^2\theta-1}}d\theta$$

$$= 2\left(\frac{\pi}{3}\right) - 1(0) - \int_{\theta=0}^{\pi/3}\frac{\sec\theta\tan\theta}{\sqrt{\tan^2\theta}}d\theta = \frac{2\pi}{3} - \int_{\theta=0}^{\pi/3}\sec\theta\, d\theta$$

$$= \frac{2\pi}{3} - [\ln|\sec\theta + \tan\theta|]_{\theta=0}^{\frac{\pi}{3}} = \frac{2\pi}{3} - \ln(2+\sqrt{3}) + \ln(1+0)$$

$$= \frac{2\pi}{3} - \ln(2+\sqrt{3}) + \ln 1 = \frac{2\pi}{3} - \ln(2+\sqrt{3}) + 0 = \frac{2\pi}{3} - \ln(2+\sqrt{3}) \approx 0.777$$

Alternate answer: $\frac{2\pi}{3} - \frac{1}{2}\ln(2+\sqrt{3}) = \frac{2\pi}{3} + \frac{1}{2}\ln(2-\sqrt{3})$ are equivalent. It's because

$$\frac{1}{2+\sqrt{3}} = \frac{1}{2+\sqrt{3}}\frac{2-\sqrt{3}}{2-\sqrt{3}} = \frac{2-\sqrt{3}}{4-2\sqrt{3}+2\sqrt{3}-3} = \frac{2-\sqrt{3}}{1} = 2-\sqrt{3}.$$ Using the rule $-\ln p = \ln p^{-1} = \ln\frac{1}{p}$,

$-\ln(2+\sqrt{3}) = \ln\left(\frac{1}{2+\sqrt{3}}\right) = \ln(2-\sqrt{3})$. If you're still not convinced, your calculator should show you that $\frac{2\pi}{3} - \ln(2+\sqrt{3})$ and $\frac{2\pi}{3} + \frac{1}{2}\ln(2-\sqrt{3})$ are both ≈ 0.777.

Check the antiderivative: The antiderivative is the expression we found before plugging in the limits: $x\sec^{-1}x - \ln(\sec\theta + \tan\theta) = x\sec^{-1}x - \ln[\sec(\sec^{-1}x) + \tan(\sec^{-1}x)] = x\sec^{-1}x - \ln[x + \sqrt{x^2-1}]$. Apply the product rule and the chain rule. Notes: Since $x = \sec\theta$, it follows that $\theta = \sec^{-1}x$. See the footnotes in Chapter 3 regarding inverse trig identities. Recall from first-semester calculus that $\frac{d}{dx}\sec^{-1}x = \frac{1}{x\sqrt{x^2-1}}$ (from the limits, the angle lies in Quadrant 1 in this problem).

$$\frac{d}{dx}\left(x\sec^{-1}x - \ln\left[x+\sqrt{x^2-1}\right]\right)$$

$$= x\frac{d}{dx}\sec^{-1}x + \sec^{-1}x\frac{d}{dx}x - \frac{1}{x+\sqrt{x^2-1}}\frac{d}{dx}\left(x+\sqrt{x^2-1}\right)$$

$$= x\left(\frac{1}{x\sqrt{1-x^2}}\right) + (\sec^{-1}x)(1) - \frac{1}{x+\sqrt{x^2-1}}\left(1+\frac{2x}{2\sqrt{x^2-1}}\right)$$

$$= \frac{1}{\sqrt{1-x^2}} + \sec^{-1}x - \frac{1}{x+\sqrt{x^2-1}}\left(\frac{\sqrt{x^2-1}}{\sqrt{x^2-1}} + \frac{x}{\sqrt{x^2-1}}\right)$$

$$= \frac{1}{\sqrt{1-x^2}} + \sec^{-1}x - \frac{1}{x+\sqrt{x^2-1}}\left(\frac{x+\sqrt{x^2-1}}{\sqrt{x^2-1}}\right)$$

$$= \frac{1}{\sqrt{1-x^2}} + \sec^{-1}x - \frac{1}{\sqrt{x^2-1}} = \sec^{-1}x$$

Answers with Full Solutions

④ $u = w, dv = \cosh w\, dw \rightarrow du = dw, v = \sinh w$

Tip: Find the derivative of u to get du, but **integrate** dv to find v.

$$\int w \cosh w\, dw = w \sinh w - \int \sinh w\, dw = w \sinh w - \cosh w + c$$

Note: This problem features **hyperbolic** cosine (Chapter 4), **not** ordinary cosine. Recall that $\int \sinh w\, dw = +\cosh w$ **differs in sign** from $\int \sin \theta\, d\theta = -\cos \theta$.

Check the answer: Apply the product rule.

$$\frac{d}{dw}(w \sinh w - \cosh w + c) = w \frac{d}{dw}\sinh w + \sinh w \frac{d}{dw}w - \sinh w$$

$$= w \cosh w + (\sinh w)(1) - \sinh w = w \cosh w$$

Note: Recall from Chapter 4 that $\frac{d}{dw}\cosh w = \sinh w$, **unlike** $\frac{d}{d\theta}\cos \theta = -\sin \theta$.

⑤ $u = \ln x, dv = x\, dx \rightarrow du = \frac{dx}{x}, v = \frac{x^2}{2}$

Remember: Find the derivative of u to get du, but **integrate** dv to find v.

$$\int x \ln x\, dx = (\ln x)\left(\frac{x^2}{2}\right) - \int \frac{x^2}{2}\frac{dx}{x} = \frac{x^2}{2}\ln x - \frac{1}{2}\int x\, dx = \frac{x^2}{2}\ln x - \frac{x^2}{4} + c$$

Alternate answers: $\frac{x^2}{2}\left[\ln(x) - \frac{1}{2}\right] + c = \frac{x^2}{4}[2\ln(x) - 1] + c$

Check the answer: Apply the product rule.

$$\frac{d}{dx}\left(\frac{x^2}{2}\ln x - \frac{x^2}{4} + c\right) = \frac{x^2}{2}\frac{d}{dx}\ln x + \ln x \frac{d}{dx}\left(\frac{x^2}{2}\right) - \frac{x}{2} = \frac{x^2}{2}\left(\frac{1}{x}\right) + (\ln x)(x) - \frac{x}{2}$$

$$= \frac{x}{2} + x \ln x - \frac{x}{2} = x \ln x$$

⑥ $u = t, dv = e^t dt \rightarrow du = dt, v = e^t$

Remember: Find the derivative of u to get du, but **integrate** dv to find v.

$$\int_{t=0}^{1} t e^t\, dt = [te^t]_{t=0}^1 - \int_{t=0}^{1} e^t\, dt = 1e^1 - 0e^0 - [e^t]_{t=0}^1 = e^1 - 0 - (e^1 - e^0)$$

$$= e^1 - e^1 + e^0 = 0 + 1 = 1$$

Notes: With a **definite integral**, remember to evaluate uv over the integration limits. Recall that a definite integral represents the area between the curve and the horizontal axis. Since te^t is positive from $t = 0$ to $t = 1$, the area under the curve must be positive. So if you got a negative answer for the integral, you know that you made a mistake somewhere (perhaps with the signs when subtracting $[e^t]_{t=0}^1$).

Check the antiderivative: The antiderivative is the expression we found before plugging in the limits: $te^t - e^t = e^t(t-1)$. Apply the product rule.

$$\frac{d}{dt}(te^t - e^t) = t\frac{d}{dt}e^t + e^t\frac{d}{dt}t - e^t = te^t + (e^t)(1) - e^t = te^t + e^t - e^t = te^t$$

❼ Note: $\tan^2 y$ means to first find $\tan y$ and then square the tangent, whereas $\tan(y^2)$ would mean to first square y before finding the tangent. Compare Exercises 7 and 9.

$u = y, dv = \tan^2 y\, dy \to du = dy,$

$$v = \int \tan^2 y\, dy = \int (\sec^2 y - 1)\, dy = \int \sec^2 y\, dy - \int dy = \tan y - y$$

Remember: Find the derivative of u to get du, but **integrate** dv to find v.

$$\int y\tan^2 y\, dy = y(\tan y - y) - \int (\tan y - y)\, dy = y\tan y - y^2 - \int \tan y\, dy + \int y\, dy$$

$$= y\tan y - y^2 - \ln|\sec y| + \frac{y^2}{2} + c = \frac{y^2}{2} - y^2 + y\tan y - \ln|\sec y| + c$$

$$= -\frac{y^2}{2} + y\tan y - \ln|\sec y| + c \quad \text{Alternate answer: } = -\frac{y^2}{2} + y\tan y + \ln|\cos y| + c$$

Check the answer: Apply the chain rule and the product rule.

$$\frac{d}{dy}\left[-\frac{y^2}{2} + y\tan y - \ln(\sec y) + c\right] = -y + y\frac{d}{dy}\tan y + \tan y\frac{d}{dy}y - \frac{1}{\sec y}\frac{d}{dy}\sec y$$

$$= -y + y\sec^2 y + (\tan y)(1) - \frac{1}{\sec y}\sec y\tan y = -y + y\sec^2 y + \tan y - \tan y$$

$$= -y + y\sec^2 y = y(\sec^2 y - 1) = y\tan^2 y$$

❽ $u = \ln x, dv = \frac{dx}{x^3} = x^{-3}dx \to du = \frac{dx}{x}, v = -\frac{1}{2x^2} = -\frac{x^{-2}}{2}$

Remember: Find the derivative of u to get du, but **integrate** dv to find v.

$$\int \frac{\ln x}{x^3}dx = (\ln x)\left(-\frac{1}{2x^2}\right) - \int\left(-\frac{1}{2x^2}\right)\frac{dx}{x} = -\frac{\ln x}{2x^2} + \frac{1}{2}\int\frac{dx}{x^3} = -\frac{\ln x}{2x^2} + \frac{1}{2}\int x^{-3}dx$$

$$= -\frac{\ln x}{2x^2} + \frac{1}{2}\frac{x^{-2}}{(-2)} + c = -\frac{\ln x}{2x^2} - \frac{1}{4x^2} + c = -\frac{1}{2x^2}\left(\frac{1}{2} + \ln x\right) + c = -\frac{1}{4x^2}(1 + 2\ln x) + c$$

Question: How is this answer negative, yet the integrand $\frac{\ln x}{x^3}$ is positive? Isn't the area under this curve positive (see the note in the solution to Exercise 6)? The area will be positive if you do a **definite** integral. Example: $\int_{x=1}^{2}\frac{\ln x}{x^3}dx = -\frac{1+2\ln 2}{4(2)^2} + \frac{1+2\ln 1}{4(1)^2} \approx$

$-0.149 + 0.25 \approx 0.101$.

Answers with Full Solutions

Check the answer: Apply the product rule.

$$\frac{d}{dx}\left(-\frac{\ln x}{2x^2} - \frac{1}{4x^2} + c\right) = -\frac{\ln x}{2}\frac{d}{dx}\frac{1}{x^2} - \frac{1}{2x^2}\frac{d}{dx}\ln x - \frac{1}{4}\frac{d}{dx}\frac{1}{x^2}$$

$$= -\frac{\ln x}{2}\frac{d}{dx}x^{-2} - \frac{1}{2x^2}\frac{1}{x} - \frac{1}{4}\frac{d}{dx}x^{-2} = -\frac{\ln x}{2}(-2)x^{-3} - \frac{1}{2x^3} - \frac{1}{4}(-2)x^{-3}$$

$$= \frac{\ln x}{x^3} - \frac{1}{2x^3} + \frac{1}{2x^3} = \frac{\ln x}{x^3}$$

❾ Note: $\sin(\varphi^2)$ means to first square φ and then take the sine, whereas $\sin^2 \varphi$ would mean to first find the sine of φ and then square the sine. Compare Exercises 7 and 9.

$u = \varphi^2, dv = \varphi \sin(\varphi^2) d\varphi \to du = 2\varphi d\varphi$,

$$v = \int \varphi \sin(\varphi^2) d\varphi = \frac{1}{2}\int \sin t \, dt = -\frac{1}{2}\cos t = -\frac{1}{2}\cos(\varphi^2)$$

Remember: Find the derivative of u to get du, but **integrate** dv to find v.

We let $t = \varphi^2$ such that $dt = 2\varphi d\varphi$ or $\frac{dt}{2} = \varphi d\varphi$ in the above integral to find v.

Notes: This problem is 'tricky.' Many students try $u = \varphi^3$ and $dv = \sin(\varphi^2)$, but then they're stuck trying to find the antiderivative of $dv = \sin(\varphi^2)$. The trick is to realize that $dv = \varphi \sin(\varphi^2) d\varphi$ is easy to integrate with the substitution $t = \varphi^2, dt = 2\varphi d\varphi$, leading to $v = -\frac{1}{2}\cos(\varphi^2)$, which means that $u = \varphi^2$ such that $udv = \varphi^3 \sin(\varphi^2) d\varphi$.

$$\int_{\varphi=0}^{\sqrt{\pi/2}} \varphi^3 \sin(\varphi^2) d\varphi = \left[-\varphi^2 \frac{1}{2}\cos(\varphi^2)\right]_{\varphi=0}^{\sqrt{\pi/2}} + \int_{\varphi=0}^{\sqrt{\pi/2}} \frac{1}{2}\cos(\varphi^2) 2\varphi \, d\varphi$$

$$= -\frac{1}{2}\left(\sqrt{\frac{\pi}{2}}\right)^2 \cos\left(\sqrt{\frac{\pi}{2}}\right)^2 + \frac{1}{2}0^2 \cos 0^2 + \int_{\varphi=0}^{\sqrt{\pi/2}} \varphi \cos(\varphi^2) d\varphi$$

$w = \varphi^2, dw = 2\varphi d\varphi, \frac{dw}{2} = \varphi d\varphi$

At $\varphi = 0, w = 0^2 = 0$. At $\varphi = \sqrt{\frac{\pi}{2}}, w = \left(\sqrt{\frac{\pi}{2}}\right)^2 = \frac{\pi}{2}$.

$$\int_{\varphi=0}^{\sqrt{\pi/2}} \varphi^3 \sin(\varphi^2) d\varphi = -\frac{1}{2}\frac{\pi}{2}\cos\left(\frac{\pi}{2}\right) + \frac{1}{2}(0)(1) + \frac{1}{2}\int_{w=0}^{\pi/2} \cos w \, dw = -\frac{\pi}{4}(0) + 0 + \frac{1}{2}[\sin w]_{w=0}^{\pi/2}$$

$$= \frac{1}{2}\sin\frac{\pi}{2} - \frac{1}{2}\sin 0 = \frac{1}{2}(1) - \frac{1}{2}(0) = \frac{1}{2} = 0.5$$

Techniques of Integration Calculus Practice Workbook

Check the antiderivative: The antiderivative is the expression we found before plugging in the limits: $-\frac{\varphi^2}{2}\cos(\varphi^2) + \frac{1}{2}\sin w = -\frac{\varphi^2}{2}\cos(\varphi^2) + \frac{1}{2}\sin(\varphi^2)$. Apply the chain rule and the product rule.

$$\frac{d}{d\varphi}\left[-\frac{\varphi^2}{2}\cos(\varphi^2) + \frac{1}{2}\sin(\varphi^2)\right] = -\frac{\varphi^2}{2}\frac{d}{d\varphi}\cos(\varphi^2) - \frac{1}{2}\cos(\varphi^2)\frac{d}{d\varphi}\varphi^2 + \frac{1}{2}\cos(\varphi^2)\frac{d}{d\varphi}\varphi^2$$

$$= \frac{\varphi^2}{2}\sin^2(\varphi)\frac{d}{d\varphi}\varphi^2 - \frac{1}{2}[\cos(\varphi^2)](2\varphi) + \frac{1}{2}[\cos(\varphi^2)](2\varphi) = \left[\frac{\varphi^2}{2}\sin^2(\varphi)\right](2\varphi) + 0$$

$$= \varphi^3 \sin^2(\varphi)$$

⑩ $u = e^x, dv = \sin x\, dx \rightarrow du = e^x dx, v = -\cos x$

Remember: Find the derivative of u to get du, but **integrate** dv to find v.

$$\int e^x \sin x\, dx = e^x(-\cos x) - \int(-\cos x)e^x\, dx = -e^x \cos x + \int e^x \cos x\, dx$$

Similar to Example 6, we need to integrate by parts a **second** time.

$u_2 = e^x, dv_2 = \cos x\, dx \rightarrow du_2 = e^x dx, v_2 = \sin x$

$$\int e^x \sin x\, dx = -e^x \cos x + e^x \sin x - \int \sin x\, e^x\, dx$$

$$\int e^x \sin x\, dx = -e^x \cos x + e^x \sin x - \int e^x \sin x\, dx$$

Add $\int e^x \sin x\, dx$ to both sides of the equation.

$$2\int e^x \sin x\, dx = -e^x \cos x + e^x \sin x$$

Divide by 2 on both sides and add the constant of integration.

$$\int e^x \sin x\, dx = -\frac{e^x}{2}\cos x + \frac{e^x}{2}\sin x + c = \frac{e^x}{2}(\sin x - \cos x) + c$$

Check the answer: Apply the product rule.

$$\frac{d}{dx}\left(-\frac{e^x}{2}\cos x + \frac{e^x}{2}\sin x\right) = -\frac{e^x}{2}\frac{d}{dx}\cos x - \frac{\cos x}{2}\frac{d}{dx}e^x + \frac{e^x}{2}\frac{d}{dx}\sin x + \frac{\sin x}{2}\frac{d}{dx}e^x$$

$$= -\frac{e^x}{2}(-\sin x) - \frac{e^x}{2}\cos x + \frac{e^x}{2}\cos x + \frac{e^x}{2}\sin x = \frac{e^x}{2}\sin x + \frac{e^x}{2}\sin x = e^x \sin x$$

⑪ $u = \ln y, dv = \frac{dy}{\sqrt{y}} = y^{-1/2}dy \rightarrow du = \frac{dy}{y}, v = 2y^{1/2} = 2\sqrt{y}$

Remember: Find the derivative of u to get du, but **integrate** dv to find v.

$$\int \frac{\ln y}{\sqrt{y}}dy = (\ln y)(2\sqrt{y}) - \int \frac{2\sqrt{y}}{y}dy = 2\sqrt{y}\ln y - 2\int \frac{dy}{\sqrt{y}} = 2\sqrt{y}\ln y - 2\int y^{-1/2}dy$$

Answers with Full Solutions

$= 2\sqrt{y}\ln y - 4y^{1/2} + c = 2\sqrt{y}\ln y - 4\sqrt{y} + c = 2\sqrt{y}(-2 + \ln y) + c$

Notes: $\frac{y}{\sqrt{y}} = \frac{y\sqrt{y}}{\sqrt{y}\sqrt{y}} = \frac{y\sqrt{y}}{y} = \sqrt{y} = y^{1/2}$ and $\frac{1}{\sqrt{y}} = \frac{1}{y^{1/2}} = y^{-1/2}$.

Check the answer: Apply the product rule.

$\frac{d}{dy}(2\sqrt{y}\ln y - 4\sqrt{y} + c) = \frac{d}{dy}(2y^{1/2}\ln y - 4y^{1/2} + c)$

$= 2y^{1/2}\frac{d}{dy}\ln y + 2\ln y\frac{d}{dy}y^{1/2} - 4\left(\frac{1}{2}\right)y^{-1/2} = 2y^{1/2}\left(\frac{1}{y}\right) + 2\ln y\left(\frac{1}{2}\right)y^{-1/2} - 2y^{-1/2}$

$= 2y^{-1/2} + y^{-1/2}\ln y - 2y^{-1/2} = \frac{\ln y}{\sqrt{y}}$

⑫ $u = \ln(\tan\theta), dv = \sec^2\theta\, d\theta \rightarrow du = \frac{\sec^2\theta}{\tan\theta}d\theta, v = \tan\theta$

Note: Use the chain rule. Let $t = \tan\theta$. Then $u = \ln t$ and

$\frac{du}{d\theta} = \frac{du}{dt}\frac{dt}{d\theta} = \frac{d}{dt}\ln t\frac{d}{d\theta}\tan\theta = \frac{1}{t}\sec^2\theta = \frac{\sec^2\theta}{\tan\theta} \rightarrow du = \frac{\sec^2\theta}{\tan\theta}d\theta$

Remember: Find the derivative of u to get du, but **integrate** dv to find v.

$\int_{\theta=\pi/4}^{\pi/3}\sec^2\theta\ln(\tan\theta)\, d\theta = [\tan\theta\ln(\tan\theta)]_{\theta=\pi/4}^{\pi/3} - \int_{\theta=\pi/4}^{\pi/3}\tan\theta\frac{\sec^2\theta}{\tan\theta}d\theta$

$= \tan\frac{\pi}{3}\ln\left(\tan\frac{\pi}{3}\right) - \tan\frac{\pi}{4}\ln\left(\tan\frac{\pi}{4}\right) - \int_{\theta=\pi/4}^{\pi/3}\sec^2\theta\, d\theta$

$= \sqrt{3}\ln\sqrt{3} - 1\ln 1 - [\tan\theta]_{\theta=\pi/4}^{\pi/3} = \sqrt{3}\ln\sqrt{3} - 1(0) - (\sqrt{3} - 1)$

$= \sqrt{3}\ln\sqrt{3} - \sqrt{3} + 1 = 1 - \sqrt{3} + \sqrt{3}\ln\sqrt{3} \approx 0.219$

Check the antiderivative: The antiderivative is the expression we found before plugging in the limits: $\tan\theta\ln(\tan\theta) - \tan\theta$. Apply the chain rule and the product rule.

$\frac{d}{d\theta}[\tan\theta\ln(\tan\theta) - \tan\theta] = \tan\theta\frac{d}{d\theta}\ln(\tan\theta) + \ln(\tan\theta)\frac{d}{d\theta}\tan\theta - \sec^2\theta$

$= \tan\theta\frac{1}{\tan\theta}\frac{d}{d\theta}\tan\theta + [\ln(\tan\theta)]\sec^2\theta - \sec^2\theta$

$= \sec^2\theta + \sec^2\theta\ln(\tan\theta) - \sec^2\theta = \sec^2\theta\ln(\tan\theta)$

⑬ $u = x^2, dv = x\sqrt{x^2 - 1}dx \rightarrow du = 2xdx,$

$v = \int x\sqrt{x^2 - 1}\, dx = \int\frac{\sqrt{w}}{2}dw = \frac{1}{2}\int w^{1/2}\, dw = \frac{w^{3/2}}{3} = \frac{(x^2 - 1)^{3/2}}{3}$

Techniques of Integration Calculus Practice Workbook

Remember: Find the derivative of u to get du, but **integrate** dv to find v. In the above integral, $w = x^2 - 1$ such that $dw = 2xdx$. This integral is just to find v.

Notes: This problem is 'tricky.' Many students try $u = x^3$ and $dv = \sqrt{x^2 - 1}$, but the antiderivative of $\sqrt{x^2 - 1}$ is not a simple expression (you could find that antiderivative with trig sub, but the answer won't be nice to work with). The trick is to realize that $dv = x\sqrt{x^2 - 1}dx$ is easy to integrate with the substitution $w = x^2 - 1$, $dw = 2xdx$, leading to $v = \frac{(x^2-1)^{3/2}}{3}$, which means that $u = x^2$ such that $udv = x^3\sqrt{x^2 - 1}dx$.

$$\int x^3\sqrt{x^2 - 1}\,dx = x^2\frac{(x^2 - 1)^{3/2}}{3} - \int \frac{(x^2 - 1)^{3/2}}{3} 2x\,dx = \frac{x^2(x^2 - 1)^{3/2}}{3} - \frac{2}{3}\int (x^2 - 1)^{3/2}x\,dx$$

$w = x^2 - 1, dw = 2xdx, \dfrac{dw}{2} = xdx$

$$= \frac{x^2(x^2 - 1)^{3/2}}{3} - \frac{2}{3}\int (x^2 - 1)^{3/2}x\,dx = \frac{x^2(x^2 - 1)^{3/2}}{3} - \frac{2}{3}\int \frac{w^{3/2}}{2}\,dw$$

$$= \frac{x^2(x^2 - 1)^{3/2}}{3} - \frac{1}{3}\int w^{3/2}\,dw = \frac{x^2(x^2 - 1)^{3/2}}{3} - \frac{1}{3}\frac{2w^{5/2}}{5} + c$$

$$= \frac{x^2(x^2 - 1)^{3/2}}{3} - \frac{2(x^2 - 1)^{5/2}}{15} + c$$

Check the answer: Apply the chain rule and the product rule.

$$\frac{d}{dx}\left[\frac{x^2(x^2 - 1)^{3/2}}{3} - \frac{2(x^2 - 1)^{5/2}}{15} + c\right]$$

$$= \frac{x^2}{3}\frac{d}{dx}(x^2 - 1)^{3/2} + \frac{(x^2 - 1)^{3/2}}{3}\frac{d}{dx}x^2 - \frac{2}{15}\left(\frac{5}{2}\right)(x^2 - 1)^{3/2}\frac{d}{dx}(x^2 - 1)$$

$$= \frac{x^2}{3}\left(\frac{3}{2}\right)(x^2 - 1)^{1/2}\frac{d}{dx}(x^2 - 1) + \frac{(x^2 - 1)^{3/2}}{3}(2x) - \frac{1}{3}(x^2 - 1)^{3/2}(2x)$$

$$= \frac{x^2}{2}\sqrt{x^2 - 1}(2x) + \frac{2x(x^2 - 1)^{3/2}}{3} - \frac{2x}{3}(x^2 - 1)^{3/2} = x^3\sqrt{x^2 - 1}$$

⑭ First let $t = \ln z$, $dt = \dfrac{dz}{z}$.

Exponentiate both sides of $t = \ln z$ to get $e^t = e^{\ln z} = z$ (Chapter 4). From $dt = \dfrac{dz}{z}$, we get $dz = zdt = e^t dt$ (using $z = e^t$). Replace $\ln z$ with t and replace dz with $e^t dt$.

$$\int \cos(\ln z)\,dz = \int \cos t\, e^t dt = \int e^t \cos t\,dt$$

$u = e^t, dv = \cos t\,dt \rightarrow du = e^t dt, v = \sin t$

Answers with Full Solutions

Remember: Find the derivative of u to get du, but **integrate** dv to find v.

$$\int e^t \cos t \, dt = e^t \sin t - \int \sin t \, e^t \, dt$$

Similar to Example 6, we need to integrate by parts a **second** time.

$u_2 = e^t, dv_2 = \sin t \, dt \to du_2 = e^t dt, v_2 = -\cos t$

$$\int e^t \cos t \, dt = e^t \sin t - \left[e^t(-\cos t) - \int (-\cos t) e^t \, dt \right]$$

$$\int e^t \cos t \, dt = e^t \sin t - \left[-e^t \cos t + \int e^t \cos t \, dt \right] = e^t \sin t + e^t \cos t - \int e^t \cos t \, dt$$

Add $\int e^x \sin x \, dx$ to both sides of the equation.

$$2 \int e^t \cos t \, dt = e^t \sin t + e^t \cos t$$

Divide by 2 on both sides and add the constant of integration.

$$\int e^t \cos t \, dt = \frac{e^t}{2} \sin t + \frac{e^t}{2} \cos t + c = \frac{z}{2} \sin(\ln z) + \frac{z}{2} \cos(\ln z) + c$$

Note: Recall the identity $e^{\ln z} = z$. (We used this in Chapter 4.)

Check the answer: Apply the chain rule and product rule.

$$\frac{d}{dt}\left(\frac{z}{2} \sin(\ln z) + \frac{z}{2} \cos(\ln z) + c \right)$$

$$= \frac{z}{2} \frac{d}{dz} \sin(\ln z) + \frac{\sin(\ln z)}{2} \frac{d}{dz} z + \frac{z}{2} \frac{d}{dz} \cos(\ln z) + \frac{\cos(\ln z)}{2} \frac{d}{dz} z$$

$$= \frac{z}{2} \cos(\ln z) \frac{d}{dz} \ln z + \frac{\sin(\ln z)}{2}(1) - \frac{z}{2} \sin(\ln z) \frac{d}{dz} \ln z + \frac{\cos(\ln z)}{2}(1)$$

$$= \frac{\cos(\ln z)}{2} + \frac{\sin(\ln z)}{2} - \frac{\sin(\ln z)}{2} + \frac{\cos(\ln z)}{2} = \cos(\ln z)$$

⑮ $\displaystyle \int_{w=0}^{1} \frac{w+1}{e^w} \, dw = \int_{w=0}^{1} (w+1) e^{-w} \, dw$

$u = w + 1, dv = e^{-w} dw \to du = dw, v = -e^{-w}$

Remember: Find the derivative of u to get du, but **integrate** dv to find v.

$$\int_{w=0}^{1} (w+1) e^{-w} \, dw = [(w+1)(-e^{-w})]_{w=0}^{1} - \int_{w=0}^{1} (-e^{-w}) \, dw$$

$$= -\left[\frac{w+1}{e^w}\right]_{w=0}^{1} + \int_{w=0}^{1} e^{-w} \, dw = -\frac{1+1}{e^1} + \frac{0+1}{e^0} + [-e^{-w}]_{w=0}^{1} = -\frac{2}{e} + \frac{1}{1} - \left[\frac{1}{e^w}\right]_{w=0}^{1}$$

202

Techniques of Integration Calculus Practice Workbook

$$= -\frac{2}{e} + 1 - \left(\frac{1}{e^1} - \frac{1}{e^0}\right) = -\frac{2}{e} + 1 - \frac{1}{e} + \frac{1}{1} = -\frac{2}{e} + 1 - \frac{1}{e} + 1 = \frac{-2-1}{e} + 2$$

$$= 2 - \frac{3}{e} \approx 0.896$$

Check the antiderivative: The antiderivative is the expression we found before plugging in the limits: $-(w+1)e^{-w} - e^{-w} = -we^{-w} - e^{-w} - e^{-w} = -we^{-w} - 2e^{-w}$. Apply the product rule.

$$\frac{d}{dw}(-we^{-w} - 2e^{-w}) = -w\frac{d}{dw}e^{-w} - e^{-w}\frac{d}{dw}w - 2(-1)e^{-w}$$

$$= -w(-e^{-w}) - (e^{-w})(1) + 2e^{-w} = we^{-w} - e^{-w} + 2e^{-w}$$

$$= we^{-w} + e^{-w} = (w+1)e^{-w} = \frac{w+1}{e^w}$$

⑯ First let $t = \sqrt{\theta}, dt = \frac{d\theta}{2\sqrt{\theta}} = \frac{d\theta}{2t}, 2tdt = d\theta$.

$$\int \sin(\sqrt{\theta})\, d\theta = \int \sin t\, (2t)dt = 2\int t \sin t\, dt$$

$u = t, dv = \sin t\, dt \rightarrow du = dt, v = -\cos t$

Remember: Find the derivative of u to get du, but **integrate** dv to find v.

$$2\int t \sin t\, dt = 2t(-\cos t) - 2\int(-\cos t)\, dt = -2t\cos t + 2\int \cos t\, dt$$

$$= -2t\cos t + 2\sin t + c = -2\sqrt{\theta}\cos(\sqrt{\theta}) + 2\sin(\sqrt{\theta}) + c$$

Check the answer: Apply the chain rule and the product rule.

$$\frac{d}{d\theta}\left[-2\sqrt{\theta}\cos(\sqrt{\theta}) + 2\sin(\sqrt{\theta}) + c\right]$$

$$= -2\sqrt{\theta}\frac{d}{d\theta}\cos(\sqrt{\theta}) - 2\cos(\sqrt{\theta})\frac{d}{d\theta}\sqrt{\theta} + 2\cos(\sqrt{\theta})\frac{d}{d\theta}\sqrt{\theta} = -2\sqrt{\theta}\frac{d}{d\theta}\cos(\sqrt{\theta}) + 0$$

$$= -2\sqrt{\theta}[-\sin(\sqrt{\theta})]\frac{d}{d\theta}\sqrt{\theta} = 2\sqrt{\theta}\sin(\sqrt{\theta})\frac{1}{2\sqrt{\theta}} = \sin(\sqrt{\theta})$$

⑰ $u = (\ln x)^2, dv = dx \rightarrow du = 2\ln x \frac{d}{dx}\ln x\, dx = \frac{2\ln x}{x}dx, v = x$

Note: Use the chain rule. Let $t = \ln x$. Then $u = t^2$ and

$$\frac{du}{dx} = \frac{du}{dt}\frac{dt}{dx} = \frac{d}{dt}t^2\frac{d}{dx}\ln x = 2t\left(\frac{1}{x}\right) = \frac{2t}{x} = \frac{2\ln x}{x} \rightarrow du = \frac{2\ln x}{x}dx$$

Remember: Find the derivative of u to get du, but **integrate** dv to find v.

$$\int (\ln x)^2\, dx = (\ln x)^2(x) - \int x \frac{2\ln x}{x}dx = x(\ln x)^2 - 2\int \ln x\, dx$$

$= x(\ln x)^2 - 2(x \ln x - x) + c = x(\ln x)^2 - 2x \ln x + 2x + c$

Recall from Exercise 1 that $\int \ln x \, dx = x \ln x - x$.

Check the answer: Apply the chain rule and the product rule. See the chain rule above for $\frac{d}{dx}(\ln x)^2 = \frac{2 \ln x}{x}$, as this will be used again below.

$\frac{d}{dx}[x(\ln x)^2 - 2x \ln x + 2x + c]$

$= x \frac{d}{dx}(\ln x)^2 + (\ln x)^2 \frac{d}{dx} x - 2x \frac{d}{dx} \ln x - 2 \ln x \frac{d}{dx} x + 2$

$= x \left(2 \ln x \frac{d}{dx} \ln x\right) + (\ln x)^2 (1) - 2x \left(\frac{1}{x}\right) - (2 \ln x)(1) + 2$

$= (2x \ln x)\left(\frac{1}{x}\right) + (\ln x)^2 - 2 - 2 \ln x + 2 = 2 \ln x + (\ln x)^2 - 2 \ln x = (\ln x)^2$

⑱ First use the trig identity $1 + \cot^2 \theta = \csc^2 \theta$ (Chapter 3).

$\int_{\theta=\pi/6}^{\pi/4} \cot^2 \theta \csc \theta \, d\theta = \int_{\theta=\pi/6}^{\pi/4} (\csc^2 \theta - 1) \csc \theta \, d\theta = \int_{\theta=\pi/6}^{\pi/4} \csc^3 \theta \, d\theta - \int_{\theta=\pi/6}^{\pi/4} \csc \theta \, d\theta$

$u = \csc \theta, \, dv = \csc^2 \theta \, d\theta \to du = -\csc \theta \cot \theta, \, v = -\cot \theta$

Remember: Find the derivative of u to get du, but **integrate** dv to find v.

$\int_{\theta=\pi/6}^{\pi/4} \cot^2 \theta \csc \theta \, d\theta = [(\csc \theta)(-\cot \theta)]_{\theta=\pi/6}^{\pi/4} - \int_{\theta=\pi/6}^{\pi/4} (-\cot \theta)(-\csc \theta \cot \theta) \, d\theta - \int_{\theta=\pi/6}^{\pi/4} \csc \theta \, d\theta$

$= -[\csc \theta \cot \theta]_{\theta=\pi/6}^{\pi/4} - \int_{\theta=\pi/6}^{\pi/4} \cot^2 \theta \csc \theta \, d\theta - [\ln|\csc \theta - \cot \theta|]_{\theta=\pi/6}^{\pi/4}$

Add $\int_{\theta=\pi/4}^{\pi/2} \cot^2 \theta \csc \theta \, d\theta$ to both sides of the equation.

$2 \int_{\theta=\pi/6}^{\pi/4} \cot^2 \theta \csc \theta \, d\theta = -[\csc \theta \cot \theta]_{\theta=\pi/6}^{\pi/4} - [\ln|\csc \theta - \cot \theta|]_{\theta=\pi/6}^{\pi/4}$

Divide by 2 on both sides.

$\int_{\theta=\pi/6}^{\pi/4} \cot^2 \theta \csc \theta \, d\theta = -\frac{1}{2}[\csc \theta \cot \theta]_{\theta=\pi/6}^{\pi/4} - \frac{1}{2}[\ln|\csc \theta - \cot \theta|]_{\theta=\pi/6}^{\pi/4}$

$= -\frac{1}{2} \csc \frac{\pi}{4} \cot \frac{\pi}{4} + \frac{1}{2} \csc \frac{\pi}{6} \cot \frac{\pi}{6} - \frac{1}{2} \ln \left|\csc \frac{\pi}{4} - \cot \frac{\pi}{4}\right| + \frac{1}{2} \ln \left|\csc \frac{\pi}{6} - \cot \frac{\pi}{6}\right|$

$$= -\frac{1}{2}\sqrt{2}(1) + \frac{1}{2}(2)\sqrt{3} - \frac{1}{2}\ln|\sqrt{2}-1| + \frac{1}{2}\ln|2-\sqrt{3}|$$

$$= -\frac{\sqrt{2}}{2} + \sqrt{3} + \frac{1}{2}\ln\left|\frac{2-\sqrt{3}}{\sqrt{2}-1}\right| \approx -0.7071 + 1.732 - 0.2178 \approx 0.807$$

Alternate answer: $-\frac{\sqrt{2}}{2} + \sqrt{3} + \frac{1}{2}\ln|2\sqrt{2}+2-\sqrt{6}-\sqrt{3}|$. This works because

$$\frac{2-\sqrt{3}}{\sqrt{2}-1} = \frac{2-\sqrt{3}}{\sqrt{2}-1}\left(\frac{\sqrt{2}+1}{\sqrt{2}+1}\right) = \frac{2\sqrt{2}+2-\sqrt{6}-\sqrt{3}}{2+\sqrt{2}-\sqrt{2}-1} = \frac{2\sqrt{2}+2-\sqrt{6}-\sqrt{3}}{1} = 2\sqrt{2}+2-\sqrt{6}-\sqrt{3}$$

Check the antiderivative: The antiderivative is the expression we found before plugging in the limits: $-\frac{1}{2}\csc\theta\cot\theta - \frac{1}{2}\ln|\csc\theta - \cot\theta|$. Apply the chain rule and the product rule.

$$\frac{d}{d\theta}\left[-\frac{1}{2}\csc\theta\cot\theta - \frac{1}{2}\ln(\csc\theta - \cot\theta)\right]$$

$$= -\frac{1}{2}\csc\theta\frac{d}{d\theta}\cot\theta - \frac{1}{2}\cot\theta\frac{d}{d\theta}\csc\theta - \frac{1}{2(\csc\theta - \cot\theta)}\frac{d}{d\theta}(\csc\theta - \cot\theta)$$

$$= -\frac{1}{2}\csc\theta(-\csc^2\theta) - \frac{1}{2}\cot\theta(-\csc\theta\cot\theta) - \frac{1}{2(\csc\theta - \cot\theta)}(-\csc\theta\cot\theta + \csc^2\theta)$$

$$= \frac{1}{2}\csc^3\theta + \frac{1}{2}\cot^2\theta\csc\theta + \frac{\csc\theta\cot\theta - \csc^2\theta}{2(\csc\theta - \cot\theta)}$$

$$= \frac{1}{2}\csc^3\theta + \frac{1}{2}\cot^2\theta\csc\theta + \frac{\csc\theta(\cot\theta - \csc\theta)}{2(\csc\theta - \cot\theta)}$$

$$= \frac{1}{2}\csc^3\theta + \frac{1}{2}\cot^2\theta\csc\theta - \frac{\csc\theta(\csc\theta - \cot\theta)}{2(\csc\theta - \cot\theta)} = \frac{1}{2}\csc^3\theta + \frac{1}{2}\cot^2\theta\csc\theta - \frac{1}{2}\csc\theta$$

$$= \frac{1}{2}\csc^3\theta - \frac{1}{2}\csc\theta + \frac{1}{2}\cot^2\theta\csc\theta = \frac{1}{2}\csc\theta(\csc^2\theta - 1) + \frac{1}{2}\cot^2\theta\csc\theta$$

$$= \frac{1}{2}\csc\theta\cot^2\theta + \frac{1}{2}\cot^2\theta\csc\theta = \cot^2\theta\csc\theta$$

⑲ You don't really need to use integration by parts for this integral. It turns out to be pretty easy if you just make the right trig substitutions. But since the instructions say to use integration by parts, we'll 'use' it.

First use the trig identity $1 + \tan^2 y = \sec^2 y$ (Chapter 3), but just once. It may seem tempting to write $\sec^4 y = (1 + \tan^2 y)^2$, but we'll write $\sec^4 y = \sec^2 y (1 + \tan^2 y)$.

$$\int \sec^4 y\, dy = \int \sec^2 y (1 + \tan^2 y)\, dy = \int \sec^2 y\, dy + \int \sec^2 y \tan^2 y\, dy$$

$$= \tan y + \int \sec^2 y \tan^2 y\, dy$$

Answers with Full Solutions

The first integral is obviously tan y, since $\frac{d}{dy}\tan y = \sec^2 y$. The second integral is easy if you let $w = \tan y$, $dw = \sec^2 y\, dy$ to get $\int w^2\, dw$, but we'll do it by parts just because that's what the instructions for this problem set say to do.

$u = \tan^2 y,\, dv = \sec^2 y\, dy \to du = 2\tan y \frac{d}{dy}\tan y = 2\tan y \sec^2 y\, dy,\, v = \tan y$

Note: Use the chain rule. Let $t = \tan y$. Then $u = t^2$ and

$\frac{du}{dy} = \frac{du}{dt}\frac{dt}{dy} = \frac{d}{dt}t^2 \frac{d}{dy}\tan y = 2t\sec^2 y = 2\tan y \sec^2 y \to du = 2\tan y \sec^2 y\, dy$

Remember: Find the derivative of u to get du, but **integrate** dv to find v.

$\int \sec^2 y \tan^2 y\, dy = (\tan^2 y)(\tan y) - \int (\tan y)\, 2\tan y \sec^2 y\, dy$

$\int \sec^2 y \tan^2 y\, dy = \tan^3 y - 2\int \sec^2 y \tan^2 y\, dy$

Add $2\int \sec^2 y \tan^2 y\, dy$ to both sides of the equation.

$3\int \sec^2 y \tan^2 y\, dy = \tan^3 y$

Divide by 3 on both sides and add the constant of integration.

$\int \sec^2 y \tan^2 y\, dy = \frac{\tan^3 y}{3} + c$

We're not done yet. We really want $\int \sec^4 y\, dy = \tan y + \int \sec^2 y \tan^2 y\, dy$, but all we found in the line above was $\int \sec^2 y \tan^2 y\, dy$, so we need to add tan y. (We had dropped the tan y to make it easier to find $\int \sec^2 y \tan^2 y\, dy$. If you try to work out $\int \sec^2 y \tan^2 y\, dy$ while keeping the tan y, you might run into some trouble.)

$\int \sec^4 y\, dy = \tan y + \int \sec^2 y \tan^2 y\, dy = \tan y + \frac{\tan^3 y}{3} + c$

Check the answer: Apply the chain rule.

$\frac{d}{dy}\left(\tan y + \frac{\tan^3 y}{3} + c\right) = \sec^2 y + \tan^2 y \frac{d}{dy}\tan y = \sec^2 y + \tan^2 y \sec^2 y$

$= \sec^2 y (1 + \tan^2 y) = \sec^4 y$

Let $f = \frac{w^3}{3} = \frac{\tan^3 y}{3}$ with $w = \tan y$. We used the chain rule above to get

$\frac{df}{dw}\frac{dw}{dy} = \frac{d}{dw}\frac{w^3}{3}\frac{d}{dy}\tan y = w^2 \sec^2 y = \tan^2 y \sec^2 y$

Chapter 8 Solutions

① $\int_{t=-7}^{7} 4t^5 \, dt = 0$ because t^5 is an odd function: $(-t)^5 = -t^5$.

For example, $(-2)^5 = -32 = -2^5$.

② $\int_{y=-2}^{2} 5y^4 \, dy = 2 \int_{y=0}^{2} 5y^4 \, dy = 2[y^5]_{y=0}^{2} = 2(2^5 - 0^5) = 2(32) = 64$

y^4 is an even function. For example, $(-2)^4 = 16 = 2^4$.

Check: $\int_{y=-2}^{2} 5y^4 \, dy = [y^5]_{y=-2}^{2} = 2^5 - (-2)^5 = 32 - (-32) = 32 + 32 = 64$

③ $\int_{x=-3}^{3} (x^7 - 4x^5 + 6x^3 - 9x) \, dx = 0$ because the integrand is an odd function.

Each term of the polynomial is an odd power of x, and there isn't a constant term, so if x is replaced by $-x$, every term will change sign.

④ $\int_{\theta=-\pi/6}^{\pi/6} \cos(2\theta) \, d\theta = 2 \int_{\theta=0}^{\pi/6} \cos(2\theta) \, d\theta = 2 \int_{\theta=0}^{\pi/3} \frac{\cos u}{2} \, du = \int_{\theta=0}^{\pi/3} \cos u \, du$

$= [\sin u]_{u=0}^{\pi/3} = \sin \frac{\pi}{3} - \sin 0 = \frac{\sqrt{3}}{2} - 0 = \frac{\sqrt{3}}{2}$ where $u = 2\theta$, $du = 2d\theta$, $\frac{du}{2} = d\theta$

$\cos(2\theta)$ is an even function. For example, $\cos\left[2\left(-\frac{\pi}{6}\right)\right] = \cos\left(-\frac{\pi}{3}\right) = \frac{1}{2} = \cos\left[2\left(\frac{\pi}{6}\right)\right]$.

Check: $\int_{\theta=-\pi/6}^{\pi/6} \cos(2\theta) \, d\theta = \frac{1}{2}[\sin(2\theta)]_{\theta=-\pi/6}^{\pi/6} = \frac{1}{2}\sin\left(2\frac{\pi}{6}\right) - \frac{1}{2}\sin\left(-2\frac{\pi}{6}\right) =$

$= \frac{1}{2}\sin\left(\frac{\pi}{3}\right) - \frac{1}{2}\sin\left(-\frac{\pi}{3}\right) = \frac{1}{2}\frac{\sqrt{3}}{2} - \frac{1}{2}\left(-\frac{\sqrt{3}}{2}\right) = \frac{\sqrt{3}}{4} + \frac{\sqrt{3}}{4} = \frac{\sqrt{3}}{2}$.

⑤ $\int_{\varphi=-\pi/3}^{\pi/3} \tan \varphi \, d\varphi = 0$ because $\tan \varphi$ is an odd function: $\tan(-\varphi) = -\tan \varphi$.

For example, $\tan\left(-\frac{\pi}{4}\right) = -1 = -\tan\frac{\pi}{4}$.

Answers with Full Solutions

6 $\int_{x=-\pi/4}^{\pi/4} x \sin x \tan x \, dx = 0$ because the integrand is an odd function.

The product of three odd functions is odd: x, $\sin x$, and $\tan x$ are all odd. For example, compare the values below:

$$\left(-\frac{\pi}{4}\right) \sin\left(-\frac{\pi}{4}\right) \tan\left(-\frac{\pi}{4}\right) = -\frac{\pi}{4}\left(-\frac{\sqrt{2}}{2}\right)(-1) = -\frac{\pi\sqrt{2}}{8}$$

$$\left(\frac{\pi}{4}\right) \sin\left(\frac{\pi}{4}\right) \tan\left(\frac{\pi}{4}\right) = \frac{\pi}{4}\left(\frac{\sqrt{2}}{2}\right)(1) = \frac{\pi\sqrt{2}}{8}$$

7 $\int_{z=-1}^{1} (z^3 - 2z)(z^5 + 3z^3)z \, dz = 0$ because the integrand is an odd function.

z, $z^3 - 2z$, and $z^5 + 3z^3$ are all odd functions and the product of three odd functions is an odd function.

8 $u = x - 2, du = dx$

At $x = -1, u = -1 - 2 = -3$. At $x = 3, u = 5 - 2 = 3$.

$$\int_{x=-1}^{5} [(x-2)^3 + x - 2] \, dx = \int_{u=-3}^{3} (u^3 + u) \, du = 0$$

After changing variables, the new integrand is an odd function of u over symmetric limits. Therefore, the integral is zero.

Chapter 9 Solutions

① $\int_{x=1}^{\infty} \dfrac{dx}{\sqrt{x}} = \lim_{t\to\infty} \int_{x=1}^{t} \dfrac{dx}{\sqrt{x}} = \lim_{t\to\infty} \int_{x=1}^{t} x^{-1/2}\, dx = \lim_{t\to\infty} \left[\dfrac{x^{-\frac{1}{2}+1}}{-\frac{1}{2}+1}\right]_{x=1}^{t} = \lim_{t\to\infty} \left[\dfrac{x^{1/2}}{1/2}\right]_{x=1}^{t}$

$= \lim_{t\to\infty} \left[2\sqrt{x}\right]_{x=1}^{t} = 2\lim_{t\to\infty}(\sqrt{t}-\sqrt{1}) = \infty$

This integral is divergent. Note: $\lim_{t\to\infty}\sqrt{t} = \infty$.

Check the antiderivative: Take a derivative of $2\sqrt{x}$ with respect to x.

$\dfrac{d}{dx} 2\sqrt{x} = 2\dfrac{d}{dx}x^{1/2} = (2)\left(\dfrac{1}{2}x^{-1/2}\right) = \dfrac{1}{x^{1/2}} = \dfrac{1}{\sqrt{x}}$

② $\int_{y=1}^{\infty} \dfrac{dy}{y^3} = \lim_{t\to\infty} \int_{y=1}^{t} \dfrac{dy}{y^3} = \lim_{t\to\infty} \int_{y=1}^{t} y^{-3}\, dy = \lim_{t\to\infty} \left[\dfrac{y^{-3+1}}{-3+1}\right]_{y=1}^{t} = \lim_{t\to\infty}\left[\dfrac{y^{-2}}{-2}\right]_{y=1}^{t}$

$= -\dfrac{1}{2}\lim_{t\to\infty}\left[\dfrac{1}{y^2}\right]_{y=1}^{t} = -\dfrac{1}{2}\lim_{t\to\infty}\left(\dfrac{1}{t^2}-\dfrac{1}{1^2}\right) = -\dfrac{1}{2}(0-1) = \left(-\dfrac{1}{2}\right)(-1) = \dfrac{1}{2} = 0.5$

This integral is convergent; it converges to one-half.

Check the antiderivative: Take a derivative of $-\dfrac{1}{2y^2}$ with respect to y.

$\dfrac{d}{dy}\left(-\dfrac{1}{2y^2}\right) = -\dfrac{1}{2}\dfrac{d}{dy}y^{-2} = \left(-\dfrac{1}{2}\right)(-2y^{-3}) = \dfrac{1}{y^3}$

③ $\int_{x=-1}^{1} \dfrac{dx}{\sqrt{x+1}} = \lim_{t\to -1^+} \int_{x=t}^{1} \dfrac{dx}{\sqrt{x+1}}$ (denominator goes to zero at $x = -1$)

$u = x+1,\, du = dx$

At $x = t,\, u = t+1$. At $x = 1,\, u = 1+1 = 2$.

$\lim_{t\to -1^+} \int_{x=t}^{1} \dfrac{dx}{\sqrt{x+1}} = \lim_{t\to -1^+} \int_{u=t+1}^{2} \dfrac{du}{\sqrt{u}} = \lim_{t\to -1^+} \int_{u=t+1}^{2} u^{-1/2}\, du = \lim_{t\to -1^+} \left[\dfrac{u^{-\frac{1}{2}+1}}{-\frac{1}{2}+1}\right]_{u=t+1}^{2}$

$= \lim_{t\to -1^+} \left[\dfrac{u^{1/2}}{1/2}\right]_{u=t+1}^{2} = \lim_{t\to -1^+} \left[2\sqrt{u}\right]_{u=t+1}^{2} = 2\lim_{t\to -1^+}\left(\sqrt{2}-\sqrt{t+1}\right) = 2(\sqrt{2}-0) = 2\sqrt{2}$

This integral is convergent; it converges to $2\sqrt{2} \approx 2.828$. Note: $\lim_{t\to -1^+}\sqrt{t+1} = \sqrt{0} = 0$.

Answers with Full Solutions

Check the antiderivative: Take a derivative of $f = 2\sqrt{u} = 2\sqrt{x+1}$ with respect to x.
Use the chain rule with $u = x + 1$.
$$\frac{df}{dx} = \frac{df}{du}\frac{du}{dx} = \frac{d}{du}2\sqrt{u}\frac{d}{dx}(x+1) = 2\frac{d}{du}u^{1/2}(1) = (2)\left(\frac{1}{2}u^{-1/2}\right) = \frac{1}{u^{1/2}} = \frac{1}{\sqrt{u}} = \frac{1}{\sqrt{x+1}}$$

4 $\displaystyle\int_{\theta=0}^{\pi/2} \sec^2\theta\, d\theta = \lim_{t\to\frac{\pi}{2}^-}\int_{\theta=0}^{t}\sec^2\theta\, d\theta \quad \left(\sec\theta \to \infty \text{ as } \theta \to \frac{\pi}{2} \text{ from the left}\right)$

$= \lim_{t\to\frac{\pi}{2}^-}[\tan\theta]_{\theta=0}^{t} = \lim_{t\to\frac{\pi}{2}^-}(\tan t - \tan 0) = \lim_{t\to\frac{\pi}{2}^-}(\tan t - 0) = \infty$

This integral is divergent. Notes: $\displaystyle\lim_{t\to\frac{\pi}{2}^-}\tan t = \infty$ (since $\tan t = \frac{\sin t}{\cos t}$ and $\cos t$

approaches zero in this limit).
Check the antiderivative: Take a derivative of $\tan\theta$ with respect to θ.
$$\frac{d}{d\theta}\tan\theta = \sec^2\theta$$

5 $\displaystyle\int_{x=-\infty}^{\infty}\frac{x}{x^2+16}dx = \lim_{t\to-\infty}\int_{x=t}^{0}\frac{x}{x^2+16}dx + \lim_{t\to\infty}\int_{x=0}^{t}\frac{x}{x^2+16}dx$

Notes: This problem is notably different from Example 5, since an x appears in the numerator. However, like Example 5, no real value of x can make the denominator zero (since x^2 is strictly nonnegative). Like Example 5, we chose $c = 0$.

$u = x^2 + 16, du = 2xdx, \dfrac{du}{2} = xdx$

At $x = t, u = t^2 + 16$. At $x = 0, u = 0^2 + 16 = 16$.
Do one integral at a time to see if both are convergent before adding them.

$\displaystyle\lim_{t\to-\infty}\int_{x=t}^{0}\frac{x}{x^2+16}dx = \lim_{t\to-\infty}\int_{u=t^2+16}^{16}\frac{1}{u}\frac{du}{2} = \frac{1}{2}\lim_{t\to-\infty}\int_{u=t^2+16}^{16}\frac{du}{u} = \frac{1}{2}\lim_{t\to-\infty}[\ln|u|]_{u=t^2+16}^{16}$

$= \dfrac{1}{2}\displaystyle\lim_{t\to-\infty}(\ln 16 - \ln|t^2+16|) = \infty$

This integral is divergent. Notes: $\displaystyle\lim_{t\to-\infty}\ln|t^2+16| = \infty$. Since the first integral diverged, it isn't necessary to check the second integral.

Check the antiderivative: Take a derivative of $f = \frac{1}{2}\ln u = \frac{1}{2}\ln(x^2+16)$ with respect to x. Use the chain rule with $u = x^2 + 16$.

210

Techniques of Integration Calculus Practice Workbook

$$\frac{df}{dx} = \frac{df}{du}\frac{du}{dx} = \frac{d}{du}\frac{1}{2}\ln u \frac{d}{dx}(x^2+16) = \frac{1}{2u}(2x) = \frac{x}{u} = \frac{x}{x^2+16}$$

⑥ $\displaystyle\int_{x=1}^{\infty}\frac{dx}{x\sqrt{x}} = \lim_{t\to\infty}\int_{x=1}^{t}\frac{dx}{x\sqrt{x}} = \lim_{t\to\infty}\int_{x=1}^{t}\frac{dx}{x^{3/2}} = \lim_{t\to\infty}\int_{x=1}^{t}x^{-3/2}\,dx = \lim_{t\to\infty}\left[\frac{x^{-\frac{3}{2}+1}}{-\frac{3}{2}+1}\right]_{x=1}^{t}$

$= \lim_{t\to\infty}\left[\frac{x^{-1/2}}{-1/2}\right]_{x=1}^{t} = -2\lim_{t\to\infty}\left[\frac{1}{x^{1/2}}\right]_{x=1}^{t} = -2\lim_{t\to\infty}\left[\frac{1}{\sqrt{x}}\right]_{x=1}^{t} = -2\lim_{t\to\infty}\left(\frac{1}{\sqrt{t}} - \frac{1}{\sqrt{1}}\right)$

$= -2(0-1) = -2(-1) = 2$

This integral is convergent; it converges to two. Note: $\displaystyle\lim_{t\to\infty}\frac{1}{\sqrt{t}} = 0$.

Check the antiderivative: Take a derivative of $-\frac{2}{\sqrt{x}}$ with respect to x.

$\displaystyle\frac{d}{dx}\left(-\frac{2}{\sqrt{x}}\right) = -2\frac{d}{dx}x^{-1/2} = (-2)\left(-\frac{1}{2}x^{-3/2}\right) = \frac{1}{x^{3/2}} = \frac{1}{x\sqrt{x}}$

⑦ $\displaystyle\int_{\varphi=0}^{\infty}\cos\varphi\,d\varphi = \lim_{t\to\infty}\int_{\varphi=0}^{t}\cos\varphi\,d\varphi = \lim_{t\to\infty}[\sin\varphi]_{\varphi=0}^{t} = \lim_{t\to\infty}(\sin t - \sin 0)$

$= \lim_{t\to\infty}\sin t - 0 = \lim_{t\to\infty}\sin t = $ does not exist

This integral is divergent. Although the integral isn't infinite, it doesn't converge to any one finite value; the sine function oscillates forever without ever approaching a single value. You might want to argue that the integral is bounded between -1 and 1, but it just doesn't satisfy the technical definition of a limit that exists because it never converges to a single finite value. Thus, the integral diverges. (Infinite series like $\sum_{n=1}^{\infty}\cos n$ are said to be divergent for similar reasoning.)

Check the antiderivative: Take a derivative of $\sin\varphi$ with respect to φ.

$\displaystyle\frac{d}{d\varphi}\sin\varphi = \cos\varphi$

⑧ $\displaystyle\int_{z=0}^{1}z\ln z\,dz = \lim_{t\to 0^+}\int_{z=t}^{1}z\ln z\,dz$

$u = \ln z, dv = z dz, du = \dfrac{dz}{z}, v = \dfrac{z^2}{2}$ (recall Chapter 7)

$\displaystyle\int_{i}^{f}u\,dv = [uv]_i^f - \int_i^f v\,du$ (recall Chapter 7)

Answers with Full Solutions

$$\lim_{t \to 0^+} \int_{z=t}^{1} z \ln z \, dz = \lim_{t \to 0^+} \left[(\ln z) \left(\frac{z^2}{2} \right) \right]_{z=t}^{1} - \lim_{t \to 0^+} \int_{z=t}^{1} \frac{z^2}{2} \frac{dz}{z}$$

$$= \frac{1}{2} \lim_{t \to 0^+} [z^2 \ln z]_{z=t}^{1} - \frac{1}{2} \lim_{t \to 0^+} \int_{z=t}^{1} z \, dz = \frac{1}{2} \lim_{t \to 0^+} (1^2 \ln 1 - t^2 \ln t) - \frac{1}{2} \lim_{t \to 0^+} \left[\frac{z^2}{2} \right]_{z=t}^{1}$$

$$= \frac{1}{2} \lim_{t \to 0^+} (0 - t^2 \ln t) - \frac{1}{2} \lim_{t \to 0^+} \left(\frac{1}{2} - \frac{t^2}{2} \right) = -\frac{1}{2} \lim_{t \to 0^+} t^2 \ln t - \frac{1}{2} \left(\frac{1}{2} - 0 \right)$$

$$= -\frac{1}{2} \lim_{t \to 0^+} t^2 \ln t - \frac{1}{4}$$

Since t^2 approaches zero while $\ln t$ approaches $-\infty$ in this limit, this is an indeterminate form. Apply l'Hôpital's rule like we did in Example 6. Note that $\frac{1}{1/t^2} = t^2$.

$$-\frac{1}{2} \lim_{t \to 0^+} t^2 \ln t - \frac{1}{4} = -\frac{1}{2} \lim_{t \to 0^+} \frac{\ln t}{\frac{1}{t^2}} - \frac{1}{4} = -\frac{1}{2} \left. \frac{\frac{d}{dt} \ln t}{\frac{d}{dt} \frac{1}{t^2}} \right|_{t \to 0^+} - \frac{1}{4} = -\frac{1}{2} \left(\left. \frac{1/t}{-2/t^3} \right|_{t \to 0^+} \right) - \frac{1}{4}$$

$$= -\frac{1}{2} \left. \left(\frac{1}{t} \div -\frac{2}{t^3} \right) \right|_{t \to 0^+} - \frac{1}{4} = -\frac{1}{2} \left. \left(\frac{1}{t} \times -\frac{t^3}{2} \right) \right|_{t \to 0^+} - \frac{1}{4} = \left. \frac{1}{4} \frac{t^2}{1} \right|_{t \to 0^+} - \frac{1}{4} = 0 - \frac{1}{4} = -\frac{1}{4}$$

This integral is convergent; it converges to negative one-fourth. (It's negative because the function $z \ln z$ lies below the z-axis in the interval $0 < z \leq 1$.)

Check the antiderivative: Take a derivative of $\frac{z^2}{2} \ln z - \frac{1}{2} \frac{z^2}{2} = \frac{z^2}{2} \ln z - \frac{z^2}{4}$ (the parts of the expression before we plugged in numbers) with respect to z. Use the **product rule**.

$$\frac{d}{dz} \left(\frac{z^2}{2} \ln z - \frac{z^2}{4} \right) = \frac{z^2}{2} \frac{d}{dz} \ln z + \frac{\ln z}{2} \frac{d}{dz} z^2 - \frac{z}{2} = \frac{z^2}{2} \frac{1}{z} + \frac{\ln z}{2} (2z) - \frac{z}{2} = \frac{z}{2} + z \ln z - \frac{z}{2}$$

$$= z \ln z$$

⑨ $\int_{y=-3}^{3} \frac{dy}{y^2 - 4} = \int_{y=-3}^{-2} \frac{dy}{y^2 - 4} + \int_{y=-2}^{2} \frac{dy}{y^2 - 4} + \int_{y=2}^{3} \frac{dy}{y^2 - 4}$

Notes: $y^2 - 4 = (y - 2)(y + 2)$ is zero when $y = 2$ or when $y = -2$, and both values lie within the interval from -3 to 3. This is similar to Example 4 except that in this exercise, there are two points ($y = -2$ and $y = 2$) in the interval instead of one. We need to do one integral from -3 to -2, one integral from -2 to 2, and one integral from 2 to 3. The formula above will only apply if all three integrals are convergent.

We'll apply the method of partial fractions (Chapter 6), although a trig substitution from Chapter 3 would also work. We'll work out the antiderivative first.

$$\frac{1}{y^2-4} = \frac{a}{y-2} + \frac{b}{y+2}$$

$1 = a(y+2) + b(y-2) = ay + 2a + by - 2b = y(a+b) + 2a - 2b$

Equate coefficients of like powers on both sides.

$0 = a + b$, $1 = 2a - 2b$ → $b = -a$

$1 = 2a - 2(-a) = 2a + 2a = 4a$ → $\frac{1}{4} = a$ → $b = -\frac{1}{4}$

$$\int \frac{dy}{y^2-4} = \frac{1}{4}\int \frac{dy}{y-2} - \frac{1}{4}\int \frac{dy}{y+2} = \frac{1}{4}\ln|y-2| - \frac{1}{4}\ln|y+2| = \frac{1}{4}\ln\left|\frac{y-2}{y+2}\right|$$

We used the identity $\ln p - \ln q = \ln\frac{p}{q}$. We'll do the definite integrals one at a time.

$$\int_{y=-3}^{-2} \frac{dy}{y^2-4} = \lim_{t\to -2^-} \int_{y=-3}^{t} \frac{dy}{y^2-4} = \lim_{t\to -2^-} \left[\frac{1}{4}\ln\left|\frac{y-2}{y+2}\right|\right]_{y=-3}^{t}$$

$$= \frac{1}{4}\lim_{t\to -2^-}\left(\ln\left|\frac{t-2}{t+2}\right| - \ln\left|\frac{-3-2}{-3+2}\right|\right) = \frac{1}{4}\lim_{t\to -2^-}\left(\ln\left|\frac{t-2}{t+2}\right| - \ln 5\right) = -\infty$$

This integral is divergent. Notes: $\lim_{t\to -2^-} \ln\left|\frac{t-2}{t+2}\right| = \lim_{t\to -2^-}(\ln|t-2| - \ln|t+2|) =$

$\ln|-4| - (-\infty) = \infty$ because $\ln 0 \to -\infty$. Since the first integral diverged, it isn't necessary to check the other integrals.

Check the antiderivative: Take a derivative of $\frac{1}{4}\ln(y-2) - \frac{1}{4}\ln(y+2)$ with respect to y. Apply the chain rule.

$$\frac{d}{dy}\left[\frac{1}{4}\ln(y-2) - \frac{1}{4}\ln(y+2)\right] = \frac{1}{4(y-2)}\frac{d}{dy}(y-2) - \frac{1}{4(y+2)}\frac{d}{dy}(y+2)$$

$$= \frac{1(1)}{4(y-2)} - \frac{1(1)}{4(y+2)} = \frac{y+2}{4(y-2)(y+2)} - \frac{y-2}{4(y+2)(y-2)} = \frac{y+2-(y-2)}{4(y^2-2y+2y-4)}$$

$$= \frac{y+2-y+2}{4(y^2-4)} = \frac{4}{4(y^2-4)} = \frac{1}{y^2-4}$$

⑩ $\int_{x=-\infty}^{\infty} xe^{-x^2} dx = \lim_{t\to -\infty}\int_{x=t}^{0} xe^{-x^2} dx + \lim_{t\to \infty}\int_{x=0}^{t} xe^{-x^2} dx$

Since the integrand remains finite for all real values of x, we may choose any real number for c; we chose c = 0 (like we did in Example 5).

Answers with Full Solutions

$u = -x^2, du = -2xdx, -\dfrac{du}{2} = xdx$

At $x = t, u = -t^2$. At $x = 0, u = -0^2 = 0$.

As usual, we'll work out the integrals one at a time and only join them together if both turn out to be convergent.

$$\lim_{t \to -\infty} \int_{u=-t^2}^{0} e^u \left(-\dfrac{du}{2}\right) = -\dfrac{1}{2} \lim_{t \to -\infty} \int_{u=-t^2}^{0} e^u \, du = -\dfrac{1}{2} \lim_{t \to -\infty} [e^u]_{u=-t^2}^{0}$$

$$= -\dfrac{1}{2} \lim_{t \to -\infty} \left(e^0 - e^{-t^2}\right) = -\dfrac{1}{2} \lim_{t \to -\infty} \left(1 - \dfrac{1}{e^{t^2}}\right) = -\dfrac{1}{2}(1 - 0) = -\dfrac{1}{2}$$

This is **not** the final answer yet; it's just the first integral. Notes: $\lim\limits_{t \to -\infty} e^{-t^2} = \lim\limits_{t \to -\infty} \dfrac{1}{e^{t^2}} = 0$ because $e^{t^2} \to e^{(-\infty)^2} = e^{\infty^2} \to \infty$ (and $\dfrac{1}{\infty} \to 0$); a more formal way to say it is that $\dfrac{1}{e^{t^2}}$ approaches zero as t grows very negative. $e^{-t^2} = \dfrac{1}{e^{t^2}}$ because $e^{-z} = \dfrac{1}{e^z}$.

$$\lim_{t \to \infty} \int_{u=0}^{-t^2} e^u \left(-\dfrac{du}{2}\right) = -\dfrac{1}{2} \lim_{t \to \infty} \int_{u=0}^{-t^2} e^u \, du = -\dfrac{1}{2} \lim_{t \to \infty} [e^u]_{u=0}^{-t^2}$$

$$= -\dfrac{1}{2} \lim_{t \to \infty} \left(e^{-t^2} - e^0\right) = -\dfrac{1}{2} \lim_{t \to \infty} \left(\dfrac{1}{e^{t^2}} - e^0\right) = -\dfrac{1}{2}(0 - 1) = -\dfrac{1}{2}(-1) = \dfrac{1}{2}$$

Now add these two convergent integrals together to obtain the final answer.

$$\int_{x=-\infty}^{\infty} xe^{-x^2} \, dx = \lim_{t \to -\infty} \int_{x=t}^{0} xe^{-x^2} \, dx + \lim_{t \to \infty} \int_{x=0}^{t} xe^{-x^2} \, dx = -\dfrac{1}{2} + \dfrac{1}{2} = 0$$

This integral is convergent; it converges to zero. The final answer (zero) should make sense if you recall from Chapter 8 that any odd function integrated over symmetric zeroes is equal to zero. The integrand, xe^{-x^2}, is an odd function (because x is odd while e^{-x^2} is even). For example, for $x = -2$ and $x = 2$, compare $(-2)e^{-(-2)^2} = -2e^{-4}$ with $2e^{-2^2} = 2e^{-4}$, which are identical except for the sign. Check the antiderivative: Take a derivative of $-\dfrac{1}{2}e^u = -\dfrac{1}{2}e^{-x^2}$ with respect to x, where $u = -x^2$. Apply the chain rule.

$$\dfrac{df}{dx} = \dfrac{df}{du}\dfrac{du}{dx} = \dfrac{d}{du}\left(-\dfrac{1}{2}e^u\right)\dfrac{d}{dx}(-x^2) = -\dfrac{1}{2}e^u(-2x) = xe^u = xe^{-x^2}$$

214

11 $\int_{w=0}^{3} \dfrac{dw}{w^2 - 3w + 2} = \int_{w=0}^{1} \dfrac{dw}{w^2 - 3w + 2} + \int_{w=1}^{2} \dfrac{dw}{w^2 - 3w + 2} + \int_{w=2}^{3} \dfrac{dw}{w^2 - 3w + 2}$

Notes: $w^2 - 3w + 2 = (w-1)(w-2)$ is zero when $w = 1$ or when $w = 2$, and both values lie within the interval from 0 to 3. This is similar to Example 4 except that in this exercise, there are two points ($w = 1$ and $w = 2$) in the interval instead of one. We need to do one integral from 0 to 1, one integral from 1 to 2, and one integral from 2 to 3. The formula above will only apply if all three integrals are convergent. We'll apply the method of partial fractions (Chapter 6), although completing the square from Chapter 5 would also work. We'll work out the antiderivative first.

$\dfrac{1}{w^2 - 3w + 2} = \dfrac{a}{w-1} + \dfrac{b}{w-2}$

$1 = a(w-2) + b(w-1) = aw - 2a + bw - b = w(a+b) - 2a - b$

Equate coefficients of like powers on both sides.

$0 = a + b$, $1 = -2a - b$ → $b = -a$

$1 = -2a - (-a) = -2a + a = -a$ → $-1 = a$ → $b = 1$

$\int \dfrac{dw}{w^2 - 3w + 2} = -\int \dfrac{dy}{w-1} + \int \dfrac{dy}{w-2} = -\ln|w-1| + \ln|w-2| = \ln\left|\dfrac{w-2}{w-1}\right|$

We used the identity $\ln p - \ln q = \ln \dfrac{p}{q}$. We'll do the definite integrals one at a time.

$\int_{w=0}^{1} \dfrac{dw}{w^2 - 3w + 2} = \lim_{t \to 1^-} \int_{w=0}^{t} \dfrac{dw}{w^2 - 3w + 2} = \lim_{t \to 1^-} \left[\ln\left|\dfrac{w-2}{w-1}\right|\right]_{w=0}^{t}$

$= \lim_{t \to 1^-}\left(\ln\left|\dfrac{t-2}{t-1}\right| - \ln\left|\dfrac{0-2}{0-1}\right|\right) = \lim_{t \to 1^-}\left(\ln\left|\dfrac{t-2}{t-1}\right| - \ln 2\right) = \infty$

This integral is divergent. Notes: $\lim_{t \to 1^-} \ln\left|\dfrac{t-2}{t-1}\right| = \lim_{t \to 1^-}(\ln|t-2| - \ln|t-1|)$

$= \ln|-1| - (-\infty) = \infty$ because $\ln 0 \to -\infty$. Since the first integral diverged, it isn't necessary to check the other integrals.

Check the antiderivative: Take a derivative of $-\ln|w-1| + \ln|w-2|$ with respect to w. Apply the chain rule.

$\dfrac{d}{dw}[-\ln|w-1| + \ln|w-2|] = -\dfrac{1}{(w-1)}\dfrac{d}{dw}(w-1) + \dfrac{1}{(w-2)}\dfrac{d}{dw}(w-2)$

Answers with Full Solutions

$$= -\frac{1(1)}{w-1} + \frac{1(1)}{w-2} = -\frac{w-2}{(w-1)(w-2)} + \frac{w-1}{(w-2)(w-1)} = \frac{-(w-2)+(w-1)}{w^2-w-2w+2}$$

$$= \frac{-w+2+w-1}{w^2-3w+2} = \frac{1}{w^2-3w+2}$$

⑫ $\int_{x=1}^{\infty} \frac{dx}{x^p} = \lim_{t\to\infty} \int_{x=1}^{t} \frac{dx}{x^p} = \lim_{t\to\infty} \int_{x=1}^{t} x^{-p}\, dx$

There are two cases to work out. If $p = 1$, $\int \frac{dx}{x}$ is a natural logarithm, and if $p \neq 1$ we get a different result (recall Chapters 1 and 4).

Case 1: $p = 1$.

$$\lim_{t\to\infty} \int_{x=1}^{t} x^{-p}\, dx = \lim_{t\to\infty} \int_{x=1}^{t} \frac{dx}{x} = \lim_{t\to\infty} [\ln|x|]_{x=1}^{t} = \lim_{t\to\infty}(\ln|t| - \ln 1) = \infty$$

The integral is divergent if $p = 1$.

Case 2: $p \neq 1$.

$$\lim_{t\to\infty} \int_{x=1}^{t} x^{-p}\, dx = \lim_{t\to\infty} \left[\frac{x^{-p+1}}{-p+1}\right]_{x=1}^{t} = \lim_{t\to\infty}\left(\frac{t^{-p+1}}{-p+1} - \frac{1^{-p+1}}{-p+1}\right) = \lim_{t\to\infty}\left(\frac{t^{-p+1}}{-p+1} - \frac{1}{-p+1}\right)$$

This integral diverges if $p < 1$ and converges to $0 - \frac{1}{-p+1} = \frac{1}{p-1}$ if $p > 1$. Why? If $p < 1$, then the exponent of t^{-p+1} will be positive and the limit of t^q where $q > 0$ is infinite as t grows infinite. If $p > 1$, then the exponent of t^{-p+1} will be negative and the limit of t^q where $q < 0$ is zero as t grows infinite. For example, if $p = 0.9$, then $t^{-p+1} = t^{0.1}$. As t grows infinite, $t^{0.1}$ also grows infinite (albeit somewhat slowly). In contrast, if $p = 1.1$, then $t^{-p+1} = t^{-0.1} = \frac{1}{t^{0.1}}$. As t grows infinite, $\frac{1}{t^{0.1}}$ goes to zero because $t^{0.1}$ grows infinite. This exercise has an important result: $\int_{x=1}^{\infty} \frac{dx}{x}$ diverges, $\int_{x=1}^{\infty} \frac{dx}{x^p}$ diverges if $p < 1$, and $\int_{x=1}^{\infty} \frac{dx}{x^p}$ converges if $p > 1$. For example, $\int_{x=1}^{\infty} \frac{dx}{\sqrt{x}}$ diverges (corresponding to $p = \frac{1}{2}$; see Exercise 1) and $\int_{x=1}^{\infty} \frac{dx}{x^2}$ converges (see Example 1).

Chapter 10 Solutions

① $\int_{x=0}^{\pi/3} \int_{y=0}^{1} e^{-y} \sec^2 x \, dx \, dy = \int_{x=0}^{\pi/3} \left(\int_{y=0}^{1} e^{-y} dy \right) \sec^2 x \, dx$

$= \int_{x=0}^{\pi/3} \left[-\frac{1}{e^y} \right]_{y=0}^{1} \sec^2 x \, dx = \int_{x=0}^{\pi/3} \left(-\frac{1}{e^1} + \frac{1}{e^0} \right) \sec^2 x \, dx$

$= \left(-\frac{1}{e} + \frac{1}{1} \right) \int_{x=0}^{\pi/3} \sec^2 x \, dx = \left(-\frac{1}{e} + 1 \right) [\tan x]_{x=0}^{\pi/3} = \left(1 - \frac{1}{e} \right) \left(\tan \frac{\pi}{3} - \tan 0 \right)$

$= \left(1 - \frac{1}{e} \right) (\sqrt{3} - 0) = \sqrt{3} \left(1 - \frac{1}{e} \right) = \sqrt{3} \left(\frac{e}{e} - \frac{1}{e} \right) = \frac{e-1}{e} \sqrt{3} \approx 1.095$

② $\int_{x=0}^{3} \int_{y=0}^{x} x^2 y^3 \, dx \, dy = \int_{x=0}^{3} x^2 \left(\int_{y=0}^{x} y^3 \, dy \right) dx = \int_{x=0}^{3} x^2 \left[\frac{y^4}{4} \right]_{y=0}^{x} dx$

$= \frac{1}{4} \int_{x=0}^{3} x^2 (x^4 - 0^4) \, dx = \frac{1}{4} \int_{x=0}^{3} x^6 \, dx = \frac{1}{4} \left[\frac{x^7}{7} \right]_{x=0}^{3} = \frac{1}{28} (3^7 - 0^7) = \frac{2187}{28} \approx 78.11$

③ $\int_{x=-y^2}^{y^2} \int_{y=1}^{2} \frac{x^2}{y^2} \, dx \, dy = \int_{y=1}^{2} \frac{1}{y^2} \left(\int_{x=-y^2}^{y^2} x^2 \, dx \right) dy = \int_{y=1}^{2} \frac{1}{y^2} \left[\frac{x^3}{3} \right]_{x=-y^2}^{y^2} dy$

$= \int_{y=1}^{2} \frac{1}{y^2} \left[\frac{(y^2)^3}{3} - \frac{(-y^2)^3}{3} \right] dy = \int_{y=1}^{2} \frac{1}{y^2} \left[\frac{y^6}{3} - \frac{(-y^6)}{3} \right] dy = \int_{y=1}^{2} \frac{1}{y^2} \left(\frac{y^6}{3} + \frac{y^6}{3} \right) dy$

$= \int_{y=1}^{2} \frac{1}{y^2} \left(\frac{2}{3} y^6 \right) dy = \frac{2}{3} \int_{y=1}^{2} y^4 \, dy = \frac{2}{3} \left[\frac{y^5}{5} \right]_{y=1}^{2} = \frac{2}{15} (2^5 - 1^5) = \frac{2}{15} (32 - 1) = \frac{62}{15} \approx 4.133$

Note: $(-y^2)^3 = (-y^2)(-y^2)(-y^2) = -y^2 y^2 y^2 = -y^6$.

④ $\int_{x=-1}^{3} \int_{y=1}^{x} (x^2 - 2xy + 6) \, dx \, dy$ Separate this into three double integrals.

$= \int_{x=-1}^{3} \int_{y=1}^{x} x^2 \, dx \, dy - \int_{x=-1}^{3} \int_{y=1}^{x} 2xy \, dx \, dy + 6 \int_{x=-1}^{3} \int_{y=1}^{x} dx \, dy$

Answers with Full Solutions

$$= \int_{x=-1}^{3} x^2 \left(\int_{y=1}^{x} dy \right) dx - \int_{x=-1}^{3} x \left(\int_{y=1}^{x} 2y\, dy \right) dx + 6 \int_{x=-1}^{3} \left(\int_{y=1}^{x} dy \right) dx$$

$$= \int_{x=-1}^{3} x^2 [y]_{y=1}^{x}\, dx - \int_{x=-1}^{3} x[y^2]_{y=1}^{x}\, dx + 6 \int_{x=-1}^{3} [y]_{y=1}^{x}\, dx$$

$$= \int_{x=-1}^{3} x^2(x-1)\, dx - \int_{x=-1}^{3} x(x^2-1^2)\, dx + 6 \int_{x=-1}^{3} (x-1)\, dx$$

$$= \int_{x=-1}^{3} (x^3 - x^2)\, dx - \int_{x=-1}^{3} (x^3 - x)\, dx + 6 \int_{x=-1}^{3} (x-1)\, dx$$

$$= \int_{x=-1}^{3} (x^3 - x^2 - x^3 + x + 6x - 6)\, dx = \int_{x=-1}^{3} (-x^2 + 7x - 6)\, dx$$

$$= -\left[\frac{x^3}{3}\right]_{x=-1}^{3} + 7\left[\frac{x^2}{2}\right]_{x=-1}^{3} - 6[x]_{x=-1}^{3} = -\frac{1}{3}[3^3 - (-1)^3] + \frac{7}{2}[3^2 - (-1)^2] - 6[3-(-1)]$$

$$= -\frac{1}{3}[27-(-1)] + \frac{7}{2}(9-1) - 6(3+1) = -\frac{(27+1)}{3} + \frac{7(8)}{2} - 6(4)$$

$$= -\frac{28}{3} + 28 - 24 = -\frac{28}{3} + 4 = -\frac{28}{3} + \frac{12}{3} = -\frac{16}{3} \approx -5.333$$

5 $$\int_{x=0}^{2} \int_{y=0}^{2x} (6x - 3y)^3\, dxdy = \int_{x=0}^{2} \left\{ \int_{y=0}^{2x} (6x - 3y)^3\, dy \right\} dx$$

$u = 6x - 3y, du = -3dy, -\dfrac{du}{3} = dy$ (treating x as if it were a constant)

Plug the y limits into $u = 6x - 3y$ to find the limits of u.
At $y = 0$, $u = 6x - 3(0) = 6x$. At $y = 2x$, $u = 6x - 3(2x) = 6x - 6x = 0$.

$$\int_{x=0}^{2} \left\{ \int_{u=6x}^{0} u^3 \left(-\frac{du}{3}\right) \right\} dx = -\frac{1}{3} \int_{x=0}^{2} \left(\int_{u=6x}^{0} u^3\, du \right) dx = -\frac{1}{3} \int_{x=0}^{2} \left[\frac{u^4}{4}\right]_{u=6x}^{0} dx$$

$$= -\frac{1}{12} \int_{x=0}^{2} [0^4 - (6x)^4]\, dx = -\frac{1}{12} \int_{x=0}^{2} (-1296x^4)\, dx = 108 \int_{x=0}^{2} x^4\, dx$$

$$= 108 \left[\frac{x^5}{5}\right]_{x=0}^{2} = \frac{108}{5}(2^5 - 0^5) = \frac{(108)(32)}{5} = \frac{3456}{5} = 691.2$$

6 $$\int_{x=\pi/6}^{\pi/2}\int_{y=0}^{1} y\cos(xy)\,dxdy = \int_{y=0}^{1}\left\{\int_{x=\pi/6}^{\pi/2} y\cos(xy)\,dx\right\}dy$$

Since all the limits are constants, you have a choice. Choose wisely. If you integrate over y first, you have an integral of the form $\int y\cos(ky)\,dy$, where k is a constant, which you could do by parts (Chapter 7). However, if you integrate over x first, you have an integral of the form $\int k\cos(kx)\,dx$, which is easier (since we treat y as if it were a constant while integrating over the independent variable x).

$u = xy, du = ydx$ (treating y as if it were a constant)

Plug the x limits into $u = xy$ to find the limits of u.

At $x = \frac{\pi}{6}, u = \frac{\pi y}{6}$. At $x = \frac{\pi}{2}, u = \frac{\pi y}{2}$.

Since $du = ydx$, the y in front of $\cos(xy)$ gets absorbed into the du.

$$\int_{y=0}^{1}\left\{\int_{u=\pi y/6}^{\pi y/2}\cos u\,du\right\}dy = \int_{y=0}^{1}[\sin u]_{u=\pi y/6}^{\pi y/2}\,dy = \int_{y=0}^{1}\left[\sin\left(\frac{\pi y}{2}\right) - \sin\left(\frac{\pi y}{6}\right)\right]dy$$

$$= \int_{y=0}^{1}\sin\left(\frac{\pi y}{2}\right)dy - \int_{y=0}^{1}\sin\left(\frac{\pi y}{6}\right)dy$$

Let $w = \frac{\pi y}{2}, dw = \frac{\pi}{2}dy, \frac{2dw}{\pi} = dy$ in the left integral, and let $z = \frac{\pi y}{6}, dz = \frac{\pi}{6}dy, \frac{6dz}{\pi} = dy$ in the left integral.

At $y = 0, w = \frac{\pi(0)}{2} = 0$ and $z = \frac{\pi(0)}{6} = 0$. At $y = 1, w = \frac{\pi(1)}{2} = \frac{\pi}{2}$ and $z = \frac{\pi(1)}{6} = \frac{\pi}{6}$.

$$\int_{w=0}^{\pi/2}\frac{2\sin w}{\pi}dw - \int_{z=0}^{\pi/6}\frac{6\sin z}{\pi}dz = \frac{2}{\pi}[-\cos w]_{w=0}^{\pi/2} - \frac{6}{\pi}[-\cos z]_{z=0}^{\pi/6}$$

$$= \frac{2}{\pi}\left(-\cos\frac{\pi}{2} + \cos 0\right) - \frac{6}{\pi}\left[-\cos\frac{\pi}{6} + \cos 0\right] = \frac{2}{\pi}(-0+1) - \frac{6}{\pi}\left(-\frac{\sqrt{3}}{2}+1\right)$$

$$= \frac{2}{\pi} + \frac{3\sqrt{3}}{\pi} - \frac{6}{\pi} = \frac{3\sqrt{3}-4}{\pi} \approx 0.381$$

7 $$\int_{x=0}^{1}\int_{y=x}^{2x} e^{8x-3y}\,dxdy = \int_{x=0}^{1}\left(\int_{y=x}^{2x} e^{8x-3y}\,dy\right)dx$$

219

Answers with Full Solutions

$u = 8x - 3y$, $du = -3dy$, $-\dfrac{du}{3} = dy$ (treating x as if it were a constant)

Plug the y limits into $u = 8x - 3y$ to find the limits of u.

At $y = x$, $u = 8x - 3x = 5x$. At $y = 2x$, $u = 8x - 3(2x) = 8x - 6x = 2x$.

$$\int_{x=0}^{1}\left(\int_{y=x}^{2x} e^{8x-3y}\, dy\right) dx = \int_{x=0}^{1}\left\{\int_{y=5x}^{2x} e^{u}\left(-\frac{du}{3}\right)\right\} dx = -\frac{1}{3}\int_{x=0}^{1}\left\{\int_{y=5x}^{2x} e^{u}\, du\right\} dx$$

$$= -\frac{1}{3}\int_{x=0}^{1} [e^{u}]_{y=5x}^{2x}\, dx = -\frac{1}{3}\int_{x=0}^{1}(e^{2x} - e^{5x})\, dx = \frac{1}{3}\int_{x=0}^{1}(e^{5x} - e^{2x})\, dx$$

$$= \frac{1}{3}\int_{x=0}^{1} e^{5x}\, dx - \frac{1}{3}\int_{x=0}^{1} e^{2x}\, dx = \frac{1}{3}\left[\frac{e^{5x}}{5}\right]_{x=0}^{1} - \frac{1}{3}\left[\frac{e^{2x}}{2}\right]_{x=0}^{1} = \frac{1}{15}(e^5 - e^0) - \frac{1}{6}(e^2 - e^0)$$

$$= \frac{e^5 - 1}{15} - \frac{e^2 - 1}{6} = \frac{2e^5 - 2}{30} - \frac{5e^2 - 5}{30} = \frac{2e^5 - 2 - (5e^2 - 5)}{30} = \frac{2e^5 - 2 - 5e^2 + 5}{30}$$

$$= \frac{2e^5 - 5e^2 + 3}{30} \approx 8.763$$

8 $\displaystyle\int_{r=0}^{3}\int_{\theta=0}^{2\pi} r^2 \sin^2\theta\, drd\theta = \int_{r=0}^{3} r^2\left(\int_{\theta=0}^{2\pi}\sin^2\theta\, d\theta\right) dr = \int_{r=0}^{3} r^2\left\{\int_{\theta=0}^{2\pi}\frac{1-\cos(2\theta)}{2}\, d\theta\right\} dr$

Recall the identity $\sin^2\theta = \dfrac{1-\cos(2\theta)}{2}$ from Chapter 3.

$$\int_{r=0}^{3} r^2\left\{\int_{\theta=0}^{2\pi}\frac{1}{2}\, d\theta - \int_{\theta=0}^{2\pi}\frac{\cos(2\theta)}{2}\, d\theta\right\} dr$$

$u = 2\theta$, $du = 2d\theta$, $\dfrac{du}{2} = d\theta$

At $\theta = 0$, $u = 2(0) = 0$. At $y = 2\pi$, $\theta = 2(2\pi) = 4\pi$.

$$\int_{r=0}^{3} r^2\left(\left[\frac{\theta}{2}\right]_{\theta=0}^{2\pi} - \int_{u=0}^{4\pi}\frac{\cos u}{2}\frac{du}{2}\right) dr = \int_{r=0}^{3} r^2\left(\frac{2\pi}{2} - \frac{0}{2} - \frac{1}{4}\int_{u=0}^{4\pi}\cos u\, du\right) dr$$

$$= \int_{r=0}^{3} r^2\left(\pi - \frac{1}{4}[\sin u]_{u=0}^{4\pi}\right) dr = \int_{r=0}^{3} r^2\left\{\pi - \frac{1}{4}[\sin(4\pi) - 0]\right\} dr$$

$$= \int_{r=0}^{3} r^2 \left[\pi - \frac{1}{4}(0)\right] dr = \pi \int_{r=0}^{3} r^2 \, dr = \pi \left[\frac{r^3}{3}\right]_{r=0}^{3} = \frac{\pi}{3}(3^3 - 0^3) = \frac{\pi 27}{3} = 9\pi \approx 28.27$$

⑨ $\displaystyle\int_{x=1}^{y}\int_{y=1}^{4} \frac{y^2 - x^2}{xy} dxdy = \int_{x=1}^{y}\int_{y=1}^{4} \frac{y^2}{xy} dxdy - \int_{x=1}^{y}\int_{y=1}^{4} \frac{x^2}{xy} dxdy$

$= \displaystyle\int_{x=1}^{y}\int_{y=1}^{4} \frac{y}{x} dxdy - \int_{x=1}^{y}\int_{y=1}^{4} \frac{x}{y} dxdy = \int_{y=1}^{4} y\left(\int_{x=1}^{y} \frac{dx}{x}\right) dy - \int_{y=1}^{4} \frac{1}{y}\left(\int_{x=1}^{y} x\, dx\right) dy$

$= \displaystyle\int_{y=1}^{4} y\, [\ln|x|]_{x=1}^{y} dy - \int_{y=1}^{4} \frac{1}{y}\left[\frac{x^2}{2}\right]_{x=1}^{y} dy = \int_{y=1}^{4} y(\ln|y| - \ln 1)dy - \frac{1}{2}\int_{y=1}^{4} \frac{1}{y}(y^2 - 1^2)dy$

(Since y varies from 1 to 4, we don't need the absolute values.)

$= \displaystyle\int_{y=1}^{4} y(\ln y - 0)dy - \frac{1}{2}\int_{y=1}^{4} y\, dy + \frac{1}{2}\int_{y=1}^{4} \frac{1}{y} dy = \int_{y=1}^{4} y \ln y\, dy - \frac{1}{2}\int_{y=1}^{4} y\, dy + \frac{1}{2}\int_{y=1}^{4} \frac{1}{y} dy$

$u = \ln y,\ du = \dfrac{dy}{y},\ dv = ydy,\ v = \dfrac{y^2}{2}$ (recall Chapter 7 for the first integral)

$= \left[(\ln y)\left(\dfrac{y^2}{2}\right)\right]_{y=1}^{4} - \displaystyle\int_{y=1}^{4} \frac{y^2}{2}\frac{dy}{y} - \frac{1}{2}\left[\frac{y^2}{2}\right]_{y=1}^{4} + \frac{1}{2}[\ln|y|]_{y=1}^{4}$

$= \left[\dfrac{y^2}{2}\ln y\right]_{y=1}^{4} - \dfrac{1}{2}\displaystyle\int_{y=1}^{4} y\, dy - \dfrac{1}{4}[y^2]_{y=1}^{4} + \dfrac{1}{2}(\ln 4 - \ln 1)$

$= \dfrac{4^2}{2}\ln 4 - \dfrac{1^2}{2}\ln 1 - \dfrac{1}{4}[y^2]_{y=1}^{4} - \dfrac{1}{4}(4^2 - 1^2) + \dfrac{1}{2}(\ln 4 - 0)$

$= 8\ln 4 - \dfrac{1}{2}(0) - \dfrac{1}{4}(4^2 - 1^2) - \dfrac{1}{4}(15) + \dfrac{1}{2}\ln 4 = \left(8 + \dfrac{1}{2}\right)\ln 4 - \dfrac{1}{4}(15) - \dfrac{1}{4}(15)$

$= \dfrac{17}{2}\ln 4 - \dfrac{15}{2} = -\dfrac{15}{2} + \dfrac{17}{2}\ln 4 = \dfrac{-15 + 17\ln 4}{2} \approx 4.284$

Note: Since $17\ln 4 = 17\ln 2^2 = 34\ln 2$, $-\dfrac{15}{2} + \dfrac{17}{2}\ln 4 = -\dfrac{15}{2} + 17\ln 2$.

⑩ $\displaystyle\int_{\theta=\pi/2}^{\pi}\int_{\varphi=0}^{2\pi} \sin\left(\dfrac{\theta}{3} - \dfrac{\varphi}{6}\right) d\theta d\varphi = \int_{\theta=\pi/2}^{\pi}\left\{\int_{\varphi=0}^{2\pi} \sin\left(\dfrac{\theta}{3} - \dfrac{\varphi}{6}\right) d\varphi\right\} d\theta$

$u = \dfrac{\theta}{3} - \dfrac{\varphi}{6}, du = -\dfrac{d\varphi}{6}, -6du = d\varphi$ (treating θ as if it were a constant)

Plug the φ limits into $u = \dfrac{\theta}{3} - \dfrac{\varphi}{6}$ to find the limits of u.

At $\varphi = 0, u = \dfrac{\theta}{3} - \dfrac{0}{6} = \dfrac{\theta}{3}$. At $\varphi = 2\pi, u = \dfrac{\theta}{3} - \dfrac{2\pi}{6} = \dfrac{\theta}{3} - \dfrac{\pi}{3}$.

$$\int_{\theta=\pi/2}^{\pi} \left(\int_{u=\theta/3}^{\theta/3-\pi/3} -6\sin u \, du \right) d\theta = \int_{\theta=\pi/2}^{\pi} [6\cos u]_{u=\theta/3}^{\theta/3-\pi/3} d\theta = 6 \int_{\theta=\pi/2}^{\pi} \left[\cos\left(\dfrac{\theta}{3} - \dfrac{\pi}{3}\right) - \cos\left(\dfrac{\theta}{3}\right) \right] d\theta$$

$$= 6 \int_{\theta=\pi/2}^{\pi} \cos\left(\dfrac{\theta}{3} - \dfrac{\pi}{3}\right) d\theta - 6 \int_{\theta=\pi/2}^{\pi} \cos\left(\dfrac{\theta}{3}\right) d\theta$$

$w = \dfrac{\theta}{3} - \dfrac{\pi}{3}, dw = \dfrac{d\theta}{3}, 3dw = d\theta; z = \dfrac{\theta}{3}, dz = \dfrac{d\theta}{3}, 3dz = d\theta$

At $\theta = \dfrac{\pi}{2}, w = \dfrac{0}{3} - \dfrac{1}{3}\dfrac{\pi}{2} = -\dfrac{\pi}{6}$ and $z = \dfrac{1}{3}\dfrac{\pi}{2} = \dfrac{\pi}{6}$. At $\theta = \pi, w = \dfrac{\pi}{3} - \dfrac{\pi}{3} = 0$ and $z = \dfrac{\pi}{3}$.

$$6\int_{w=-\pi/6}^{0} 3\cos w \, dw - 6\int_{z=\pi/6}^{\pi/3} 3\cos z \, d\theta = 6[3\sin w]_{w=-\pi/6}^{0} - 6[3\sin z]_{z=\pi/6}^{\pi/3}$$

$$= 18\sin 0 - 18\sin\left(-\dfrac{\pi}{6}\right) - 18\sin\left(\dfrac{\pi}{3}\right) + 18\sin\left(\dfrac{\pi}{6}\right) = 0 - 18\left(-\dfrac{1}{2}\right) - 18\left(\dfrac{\sqrt{3}}{2}\right) + 18\left(\dfrac{1}{2}\right)$$

$$= 9 - 9\sqrt{3} + 9 = 18 - 9\sqrt{3} = 9(2 - \sqrt{3}) \approx 2.412$$

Alternate solution: Use the identity $\sin(\alpha - \beta) = \sin\alpha\cos\beta - \sin\beta\cos\alpha$.

$$\int_{\theta=\pi/2}^{\pi} \int_{\varphi=0}^{2\pi} \sin\left(\dfrac{\theta}{3} - \dfrac{\varphi}{6}\right) d\theta \, d\varphi = \int_{\theta=\pi/2}^{\pi} \int_{\varphi=0}^{2\pi} \left(\sin\dfrac{\theta}{3}\cos\dfrac{\varphi}{6} - \sin\dfrac{\varphi}{6}\cos\dfrac{\theta}{3} \right) d\theta \, d\varphi$$

$$= \int_{\theta=\pi/2}^{\pi} \int_{\varphi=0}^{2\pi} \sin\dfrac{\theta}{3}\cos\dfrac{\varphi}{6} d\theta \, d\varphi - \int_{\theta=\pi/2}^{\pi} \int_{\varphi=0}^{2\pi} \sin\dfrac{\varphi}{6}\cos\dfrac{\theta}{3} d\theta \, d\varphi$$

$$= \int_{\theta=\pi/2}^{\pi} \sin\dfrac{\theta}{3} \left(\int_{\varphi=0}^{2\pi} \cos\dfrac{\varphi}{6} d\varphi \right) d\theta - \int_{\theta=\pi/2}^{\pi} \cos\dfrac{\theta}{3} \left(\int_{\varphi=0}^{2\pi} \sin\dfrac{\varphi}{6} d\varphi \right) d\theta$$

$t = \dfrac{\theta}{3}, dt = \dfrac{d\theta}{3}, 3dt = d\theta; u = \dfrac{\varphi}{6}, du = \dfrac{d\varphi}{6}, 6du = d\varphi$

At $\theta = \dfrac{\pi}{2}, t = \dfrac{1}{3}\dfrac{\pi}{2} = \dfrac{\pi}{6}$. At $\theta = \pi, t = \dfrac{\pi}{3}$. At $\varphi = 0, u = \dfrac{0}{6} = 0$. At $\varphi = 2\pi, u = \dfrac{2\pi}{6} = \dfrac{\pi}{3}$.

$$\int_{t=\pi/6}^{\pi/3} \sin t \left(\int_{u=0}^{\pi/3} 6\cos u \, du \right) 3dt - \int_{t=\pi/6}^{\pi/3} \cos t \left(\int_{u=0}^{\pi/3} 6\sin u \, du \right) 3dt$$

$$= 3 \int_{t=\pi/6}^{\pi/3} \sin t \, [6\sin u]_{u=0}^{\pi/3} \, dt - 3 \int_{t=\pi/6}^{\pi/3} \cos t \, [-6\cos u]_{u=0}^{\pi/3} \, dt$$

$$= 3 \int_{t=\pi/6}^{\pi/3} 6\sin t \left(\sin\frac{\pi}{3} - \sin 0\right) dt - 3 \int_{t=\pi/6}^{\pi/3} 6\cos t \left(-\cos\frac{\pi}{3} + \cos 0\right) dt$$

$$= 18 \int_{t=\pi/6}^{\pi/3} \sin t \left(\frac{\sqrt{3}}{2} - 0\right) dt - 18 \int_{t=\pi/6}^{\pi/3} \cos t \left(-\frac{1}{2} + 1\right) dt$$

$$= 9\sqrt{3} \int_{t=\pi/6}^{\pi/3} \sin t \, dt - 9 \int_{t=\pi/6}^{\pi/3} \cos t \, dt = 9\sqrt{3}[-\cos t]_{t=\pi/6}^{\pi/3} - 9[\sin t]_{t=\pi/6}^{\pi/3}$$

$$= 9\sqrt{3}\left(-\cos\frac{\pi}{3} + \cos\frac{\pi}{6}\right) - 9\left(\sin\frac{\pi}{3} - \sin\frac{\pi}{6}\right) = 9\sqrt{3}\left(-\frac{1}{2} + \frac{\sqrt{3}}{2}\right) - 9\left(\frac{\sqrt{3}}{2} - \frac{1}{2}\right)$$

$$= -\frac{9\sqrt{3}}{2} + \frac{9}{2}(3) - \frac{9\sqrt{3}}{2} + \frac{9}{2} = -\frac{18\sqrt{3}}{2} + \frac{27}{2} + \frac{9}{2} = -\frac{18\sqrt{3}}{2} + \frac{36}{2}$$

$$= -9\sqrt{3} + 18 = 18 - 9\sqrt{3} = 9\left(2 - \sqrt{3}\right) \approx 2.412$$

11 $\displaystyle\int_{x=2}^{y}\int_{y=3}^{4} \frac{x-y+1}{xy-2x-y+2} dxdy = \int_{x=2}^{y}\int_{y=3}^{4} \frac{x-y+1}{(x-1)(y-2)} dxdy$

Note that $(x-1)(y-2) = xy - 2x - y + 2$.

Use the method of partial fractions from Chapter 6.

$$\frac{x-y+1}{(x-1)(y-2)} = \frac{a}{x-1} + \frac{b}{y-2}$$

Multiply both sides by $(x-1)(y-2)$.

$$x - y + 1 = a(y-2) + b(x-1) = ay - 2a + bx - b$$

Observe that both sides of this equation would be true for any value of x and y if $b = 1$ and $a = -1$. To see this, plug $b = 1$ and $a = -1$ into $ay - 2a + bx - b$:

$$-1y - 2(-1) + 1x - 1 = -y + 2 + x - 1 = x - y + 1$$

Answers with Full Solutions

This means that
$$\frac{x-y+1}{(x-1)(y-2)} = \frac{a}{x-1} + \frac{b}{y-2} = \frac{-1}{x-1} + \frac{1}{y-2}$$

$$\int_{x=2}^{y} \int_{y=3}^{4} \frac{x-y+1}{(x-1)(y-2)} dx dy = \int_{x=2}^{y} \int_{y=3}^{4} \frac{-1}{x-1} dx dy + \int_{x=2}^{y} \int_{y=3}^{4} \frac{1}{y-2} dx dy$$

$$= -\int_{y=3}^{4} \left(\int_{x=2}^{y} \frac{dx}{x-1} \right) dy + \int_{y=3}^{4} \frac{1}{y-2} \left(\int_{x=2}^{y} dx \right) dy$$

$$= -\int_{y=3}^{4} [\ln|x-1|]_{x=2}^{y} \, dy + \int_{y=3}^{4} \frac{1}{y-2} [x]_{x=2}^{y} \, dy$$

$$= -\int_{y=3}^{4} [\ln(y-1) - \ln(2-1)] \, dy + \int_{y=3}^{4} \frac{1}{y-2}(y-2) \, dy$$

(We don't need the absolute values, since $y > 1$ in this interval. Also, we don't need to worry about division by zero on the right since $y > 2$ in this interval.)

$$= -\int_{y=3}^{4} [\ln(y-1) - \ln 1] \, dy + \int_{y=3}^{4} dy = -\int_{y=3}^{4} [\ln(y-1) - 0] \, dy + [y]_{y=3}^{4}$$

$$= -\int_{y=3}^{4} \ln(y-1) \, dy + (4-3) = 1 - \int_{y=3}^{4} \ln(y-1) \, dy$$

Use integration by parts (Chapter 7) with $u = \ln(y-1), du = \frac{dy}{y-1}, dv = dy, v = y$.

$$1 - \int_{y=3}^{4} \ln(y-1) \, dy = 1 - [y \ln(y-1)]_{y=3}^{4} + \int_{y=3}^{4} y \frac{dy}{y-1}$$

$$= 1 - (4 \ln 3 - 3 \ln 2) + \int_{y=3}^{4} \frac{y}{y-1} dy$$

$w = y - 1, dw = dy, y = w + 1$

At $y = 3, w = 3 - 1 = 2$. At $y = 4, w = 4 - 1 = 3$.

Replace $y - 1$ with w, replace dy with dw, and replace y with $w + 1$.

$$1 - 4\ln 3 + 3\ln 2 + \int_{w=2}^{3} \frac{w+1}{w} dw = 1 - 4\ln 3 + 3\ln 2 + \int_{w=2}^{3} \left(\frac{w}{w} + \frac{1}{w}\right) dw$$

$$= 1 - 4\ln 3 + 3\ln 2 + \int_{w=2}^{3} \left(1 + \frac{1}{w}\right) dw = 1 - 4\ln 3 + 3\ln 2 + \int_{w=2}^{3} dw + \int_{w=2}^{3} \frac{dw}{w}$$

$$= 1 - 4\ln 3 + 3\ln 2 + [w]_{w=2}^{3} + [\ln|w|]_{w=2}^{3}$$
$$= 1 - 4\ln 3 + 3\ln 2 + 3 - 2 + \ln 3 - \ln 2 = 1 + 3 - 2 - 3\ln 3 + 2\ln 2$$
$$= 2 - 3\ln 3 + 2\ln 2 = 2(1 + \ln 2) - 3\ln 3 \approx 0.0905$$

12 $\displaystyle\int_{x=1}^{2}\int_{y=0}^{x} \frac{xy}{\sqrt{x^2+y^2}} dx\, dy = \int_{x=1}^{2} x\left(\int_{y=0}^{x} \frac{y}{\sqrt{x^2+y^2}} dy\right) dx$

Integrating over y first, for the y integral, we treat x as if it were a constant, which allows us to bring the x in the numerator out of the y integral. It isn't necessary to make a trig substitution, since there is a convenient y in the numerator.

$u = x^2 + y^2, du = 2y\,dy, \dfrac{du}{2} = y\,dy$ (treating x as if it were a constant)

Plug the y limits into $u = x^2 + y^2$ to find the limits of u.
At $y = 0$, $u = x^2 + 0^2 = x^2$. At $y = x$, $u = x^2 + x^2 = 2x^2$.

$$\int_{x=1}^{2} x \left(\int_{u=x^2}^{2x^2} \frac{du}{2\sqrt{u}}\right) dx = \frac{1}{2}\int_{x=1}^{2} x \left(\int_{u=x^2}^{2x^2} u^{-1/2} du\right) dx = \frac{1}{2}\int_{x=1}^{2} x[2u^{1/2}]_{u=x^2}^{2x^2} dx$$

$$= \int_{x=1}^{2} x[\sqrt{u}]_{u=x^2}^{2x^2} dx = \int_{x=1}^{2} x\left(\sqrt{2x^2} - \sqrt{x^2}\right) dx = \int_{x=1}^{2} x(x\sqrt{2} - x) dx$$

$$= \int_{x=1}^{2} x^2(\sqrt{2} - 1) dx = (\sqrt{2} - 1) \int_{x=1}^{2} x^2 dx = (\sqrt{2} - 1)\left[\frac{x^3}{3}\right]_{x=1}^{2}$$

$$= \frac{\sqrt{2} - 1}{3}(2^3 - 1^3) = \frac{7}{3}(\sqrt{2} - 1) \approx 0.966$$

13 $\displaystyle\int_{x=0}^{1}\int_{y=0}^{2}\int_{z=0}^{3} x^3 y^2 z\, dx\,dy\,dz = \int_{x=0}^{1} x^3 \int_{y=0}^{2} y^2 \left(\int_{z=0}^{3} z\, dz\right) dx\, dy$

225

$$= \int_{x=0}^{1} x^3 \int_{y=0}^{2} y^2 \left[\frac{z^2}{2}\right]_{z=0}^{3} dxdy = \frac{1}{2} \int_{x=0}^{1} x^3 \int_{y=0}^{2} y^2 (3^2 - 0^2) \, dxdy$$

$$= \frac{9}{2} \int_{x=0}^{1} x^3 \left(\int_{y=0}^{2} y^2 \, dy\right) dx = \frac{9}{2} \int_{x=0}^{1} x^3 \left[\frac{y^3}{3}\right]_{y=0}^{2} dx = \frac{3}{2} \int_{x=0}^{1} x^3 (2^3 - 0^3) \, dx$$

$$= \frac{3}{2}(8) \int_{x=0}^{1} x^3 \, dx = 12 \left[\frac{x^4}{4}\right]_{x=0}^{1} = 3(1^4 - 0^4) = 3$$

14 $\displaystyle\int_{x=0}^{z} \int_{y=0}^{x} \int_{z=-1}^{1} (z^2 - xy) \, dxdydz = \int_{z=-1}^{1} \int_{x=0}^{z} \left\{\int_{y=0}^{x} (z^2 - xy) \, dy\right\} dxdz$

$$= \int_{z=-1}^{1} \int_{x=0}^{z} \left\{\int_{y=0}^{x} z^2 \, dy\right\} dxdz - \int_{z=-1}^{1} \int_{x=0}^{z} \left\{\int_{y=0}^{x} xy \, dy\right\} dxdz$$

$$= \int_{z=-1}^{1} z^2 \int_{x=0}^{z} \left\{\int_{y=0}^{x} dy\right\} dxdz - \int_{z=-1}^{1} \int_{x=0}^{z} x \left\{\int_{y=0}^{x} y \, dy\right\} dxdz$$

$$= \int_{z=-1}^{1} z^2 \int_{x=0}^{z} [y]_{y=0}^{x} \, dxdz - \int_{z=-1}^{1} \int_{x=0}^{z} x \left[\frac{y^2}{2}\right]_{y=0}^{x} dxdz$$

$$= \int_{z=-1}^{1} z^2 \int_{x=0}^{z} (x - 0) \, dxdz - \frac{1}{2} \int_{z=-1}^{1} \int_{x=0}^{z} x(x^2 - 0^2) \, dxdz$$

$$= \int_{z=-1}^{1} z^2 \left(\int_{x=0}^{z} x \, dx\right) dz - \frac{1}{2} \int_{z=-1}^{1} \left(\int_{x=0}^{z} x^3 \, dx\right) dz = \int_{z=-1}^{1} z^2 \left[\frac{x^2}{2}\right]_{x=0}^{z} dz - \frac{1}{2} \int_{z=-1}^{1} \left[\frac{x^4}{4}\right]_{x=0}^{z} dz$$

$$= \frac{1}{2} \int_{z=-1}^{1} z^2 (z^2 - 0^2) \, dz - \frac{1}{8} \int_{z=-1}^{1} (z^4 - 0^4) \, dz = \frac{1}{2} \int_{z=-1}^{1} z^4 \, dz - \frac{1}{8} \int_{z=-1}^{1} z^4 \, dz$$

$$= \frac{1}{2}\left[\frac{z^5}{5}\right]_{z=-1}^{1} - \frac{1}{8}\left[\frac{z^5}{5}\right]_{z=-1}^{1} = \frac{1}{10}[1^5 - (-1)^5] - \frac{1}{40}[1^5 - (-1)^5]$$

$$= \frac{4}{40}[1 - (-1)] - \frac{1}{40}[1 - (-1)] = \frac{4}{40}(1+1) - \frac{1}{40}(1+1) = \frac{8}{40} - \frac{2}{40} = \frac{6}{40} = \frac{3}{20} = 0.15$$

15 $\displaystyle\int_{x=0}^{1}\int_{y=1}^{1-x}\int_{z=1}^{y}\frac{xy}{z^2}dxdydz = \int_{x=0}^{1}x\int_{y=1}^{1-x}y\left(\int_{z=1}^{y}\frac{dz}{z^2}\right)dxdy = \int_{x=0}^{1}x\int_{y=1}^{1-x}y\left[-\frac{1}{z}\right]_{z=1}^{y}dxdy$

$= \displaystyle\int_{x=0}^{1}x\int_{y=1}^{1-x}y\left(-\frac{1}{y}+\frac{1}{1}\right)dxdy = \int_{x=0}^{1}x\int_{y=1}^{1-x}y\left(1-\frac{1}{y}\right)dxdy = \int_{x=0}^{1}x\left\{\int_{y=1}^{1-x}(y-1)\,dy\right\}dx$

$= \displaystyle\int_{x=0}^{1}x\left[\frac{y^2}{2}-y\right]_{y=1}^{1-x}dx = \int_{x=0}^{1}x\left[\frac{(1-x)^2}{2}-(1-x)-\frac{1^2}{2}+1\right]dx$

$= \displaystyle\int_{x=0}^{1}x\left(\frac{1-2x+x^2}{2}-1+x-\frac{1}{2}+1\right)dx = \int_{x=0}^{1}x\left(\frac{1}{2}-x+\frac{x^2}{2}+x-\frac{1}{2}\right)dx$

$= \displaystyle\int_{x=0}^{1}x\left(\frac{x^2}{2}\right)dx = \frac{1}{2}\int_{x=0}^{1}x^3\,dx = \frac{1}{2}\left[\frac{x^4}{4}\right]_{x=0}^{1} = \frac{1}{8}(1^4-0^4) = \frac{1}{8} = 0.125$

16 $\displaystyle\int_{x=-z}^{z}\int_{y=1}^{4}\int_{z=0}^{\sqrt{y}}\frac{dxdydz}{y} = \int_{y=1}^{4}\frac{1}{y}\int_{z=0}^{\sqrt{y}}\left(\int_{x=-z}^{z}dx\right)dzdy = \int_{y=1}^{4}\frac{1}{y}\int_{z=0}^{\sqrt{y}}[x]_{x=-z}^{z}\,dzdy$

$= \displaystyle\int_{y=1}^{4}\frac{1}{y}\int_{z=0}^{\sqrt{y}}[z-(-z)]\,dzdy = \int_{y=1}^{4}\frac{1}{y}\left(\int_{z=0}^{\sqrt{y}}2z\,dz\right)dy = 2\int_{y=1}^{4}\frac{1}{y}\left[\frac{z^2}{2}\right]_{z=0}^{\sqrt{y}}dy$

$= 2\displaystyle\int_{y=1}^{4}\frac{1}{y}\left[\frac{(\sqrt{y})^2}{2}-\frac{0^2}{2}\right]dy = 2\int_{y=1}^{4}\frac{1}{y}\left(\frac{y}{2}\right)dy = \int_{y=1}^{4}dy = [y]_{y=1}^{4} = 4-1 = 3$

17 $\displaystyle\int_{r=0}^{3}\int_{\varphi=0}^{\pi}\int_{\theta=0}^{2\pi}r^2\sin\varphi\,drd\varphi d\theta = \int_{r=0}^{3}r^2\int_{\varphi=0}^{\pi}\sin\varphi\left(\int_{\theta=0}^{2\pi}d\theta\right)d\varphi dr$

$= \displaystyle\int_{r=0}^{3}r^2\int_{\varphi=0}^{\pi}\sin\varphi\,[\theta]_{\theta=0}^{2\pi}\,d\varphi dr = \int_{r=0}^{3}r^2\int_{\varphi=0}^{\pi}\sin\varphi\,(2\pi-0)\,d\varphi dr$

$= 2\pi\displaystyle\int_{r=0}^{3}r^2\left(\int_{\varphi=0}^{\pi}\sin\varphi\,d\varphi\right)dr = 2\pi\int_{r=0}^{3}r^2[-\cos\varphi]_{\varphi=0}^{\pi}\,dr$

$$= 2\pi \int_{r=0}^{3} r^2(-\cos\pi + \cos 0)\, dr = 2\pi \int_{r=0}^{3} r^2[-(-1) + 1]\, dr = 2\pi \int_{r=0}^{3} 2r^2\, dr$$

$$= 4\pi \left[\frac{r^3}{3}\right]_{r=0}^{3} = \frac{4\pi}{3}(3^3 - 0^3) = \frac{4\pi(27)}{3} = 36\pi \approx 113$$

18 $\int_{r=0}^{3}\int_{\omega=1}^{2}\int_{t=0}^{\pi/6} r^2 \omega \cos(\omega t)\, dr\, d\omega\, dt = \int_{r=0}^{3} r^2 \int_{\omega=1}^{2}\left\{\int_{t=0}^{\pi/6} \omega \cos(\omega t)\, dt\right\} d\omega\, dr$

(It's easier to integrate t before integrating ω than in the opposite order.)

$u = \omega t, du = \omega dt$ (treating ω as if it were a constant)

Plug the t limits into $u = \omega t$ to find the limits of u.

At $t = 0, u = \omega 0 = 0$. At $t = \frac{\pi}{6}, u = \omega\frac{\pi}{6} = \frac{\pi\omega}{6}$.

$$\int_{r=0}^{3} r^2 \int_{\omega=1}^{2}\left\{\int_{u=0}^{\pi\omega/6} \cos u\, du\right\} d\omega\, dr = \int_{r=0}^{3} r^2 \int_{\omega=1}^{2} [\sin u]_{u=0}^{\pi\omega/6}\, d\omega\, dr$$

$$= \int_{r=0}^{3} r^2 \int_{\omega=1}^{2}\left[\sin\left(\frac{\pi\omega}{6}\right) - \sin 0\right] d\omega\, dr = \int_{r=0}^{3} r^2\left\{\int_{\omega=1}^{2} \sin\left(\frac{\pi\omega}{6}\right) d\omega\right\} dr$$

$z = \frac{\pi\omega}{6}, du = \frac{\pi}{6} d\omega, \frac{6du}{\pi} = d\omega$

At $\omega = 1, z = \frac{\pi(1)}{6} = \frac{\pi}{6}$. At $\omega = 2, z = \frac{\pi(2)}{6} = \frac{\pi}{3}$.

$$\int_{r=0}^{3} r^2\left\{\int_{z=\pi/6}^{\pi/3} \sin z\, \frac{6dz}{\pi}\right\} dr = \frac{6}{\pi}\int_{r=0}^{3} r^2\left\{\int_{z=\pi/6}^{\pi/3} \sin z\, dz\right\} dr = \frac{6}{\pi}\int_{r=0}^{3} r^2[-\cos z]_{z=\pi/6}^{\pi/3}\, dr$$

$$= \frac{6}{\pi}\int_{r=0}^{3} r^2\left(-\cos\frac{\pi}{3} + \cos\frac{\pi}{6}\right) dr = \frac{6}{\pi}\int_{r=0}^{3} r^2\left(-\frac{1}{2} + \frac{\sqrt{3}}{2}\right) dr$$

$$= \frac{3}{\pi}(-1 + \sqrt{3})\int_{r=0}^{3} r^2\, dr = \frac{3}{\pi}(-1 + \sqrt{3})\left[\frac{r^3}{3}\right]_{r=0}^{3} = \frac{1}{\pi}(-1 + \sqrt{3})(3^3 - 0^3)$$

$$= \frac{27}{\pi}(-1 + \sqrt{3}) \approx 6.292$$

19 $\int_{x=0}^{y} \int_{y=0}^{1} \int_{z=0}^{x} xyze^{z^2} \, dxdydz = \int_{y=0}^{1} y \int_{x=0}^{y} x \left(\int_{z=0}^{x} ze^{z^2} \, dz \right) dxdy$

$u = z^2, du = 2zdz, \dfrac{du}{2} = zdz$

At $z = 0, u = 0^2 = 0$. At $z = x, u = x^2$.

$\int_{y=0}^{1} y \int_{x=0}^{y} x \left(\int_{u=0}^{x^2} e^u \dfrac{du}{2} \right) dxdy = \dfrac{1}{2} \int_{y=0}^{1} y \int_{x=0}^{y} x[e^u]_{u=0}^{x^2} \, dxdy$

$= \dfrac{1}{2} \int_{y=0}^{1} y \left\{ \int_{x=0}^{y} x(e^{x^2} - e^0) \, dx \right\} dy = \dfrac{1}{2} \int_{y=0}^{1} y \left\{ \int_{x=0}^{y} x(e^{x^2} - 1) \, dx \right\} dy$

$= \dfrac{1}{2} \int_{y=0}^{1} y \left\{ \int_{x=0}^{y} xe^{x^2} \, dx \right\} dy - \dfrac{1}{2} \int_{y=0}^{1} y \left\{ \int_{x=0}^{y} x \, dx \right\} dy$

$t = x^2, dt = 2xdx, \dfrac{dt}{2} = xdx$

At $x = 0, t = 0^2 = 0$. At $x = y, t = y^2$.

$\dfrac{1}{2} \int_{y=0}^{1} y \left(\int_{t=0}^{y^2} e^t \dfrac{dt}{2} \right) dy - \dfrac{1}{2} \int_{y=0}^{1} y \left[\dfrac{x^2}{2} \right]_{x=0}^{y} dy = \dfrac{1}{4} \int_{y=0}^{1} y[e^t]_{t=0}^{y^2} \, dy - \dfrac{1}{4} \int_{y=0}^{1} y(y^2 - 0^2) \, dy$

$= \dfrac{1}{4} \int_{y=0}^{1} y(e^{y^2} - e^0) \, dy - \dfrac{1}{4} \int_{y=0}^{1} y^3 \, dy = \dfrac{1}{4} \int_{y=0}^{1} y(e^{y^2} - 1) \, dy - \dfrac{1}{4} \left[\dfrac{y^4}{4} \right]_{y=0}^{1}$

$= \dfrac{1}{4} \int_{y=0}^{1} ye^{y^2} \, dy - \dfrac{1}{4} \int_{y=0}^{1} y \, dy - \dfrac{1}{16}(1^4 - 0^4)$

$w = y^2, dw = 2ydy, \dfrac{dw}{2} = ydy$

At $y = 0, w = 0^2 = 0$. At $y = 1, w = 1^2 = 1$.

$\dfrac{1}{4} \int_{w=0}^{1} e^w \dfrac{dw}{2} - \dfrac{1}{4} \left[\dfrac{y^2}{2} \right]_{y=0}^{1} - \dfrac{1}{16} = \dfrac{1}{8}[e^w]_{w=0}^{1} - \dfrac{1}{8}(1^2 - 0^2) - \dfrac{1}{16}$

$= \dfrac{1}{8}(e^1 - e^0) - \dfrac{1}{8} - \dfrac{1}{16} = \dfrac{e}{8} - \dfrac{1}{8} - \dfrac{1}{8} - \dfrac{1}{16} = \dfrac{2e}{16} - \dfrac{2}{16} - \dfrac{2}{16} - \dfrac{1}{16} = \dfrac{2e - 5}{16} \approx 0.0273$

WAS THIS BOOK HELPFUL?

Much effort and thought were put into this book, such as:
- Which techniques of integration to include in this workbook. Not everyone agrees on what is 'essential.' One way to interpret 'essential' is to consider which skills would be practical in applications like physics or engineering.
- Introducing the main ideas at the beginning of each chapter. The goal was to be concise while also covering the pertinent information.
- Solving examples step by step to serve as a helpful guide.
- Working out the full solutions in enough detail that students could follow along (and learn from any mistakes that they may have made).

If you appreciate the effort that went into making this book possible, there is a simple way that you could show it:

<u>Please take a moment to post an honest review.</u>

For example, you can review this book at Amazon.com or Goodreads.com.

Even a short review can be helpful and will be much appreciated. If you are not sure what to write, following are a few ideas. Better yet, write what is important to you.
- Did it help to have the full solutions to the exercises at the back of the book? Do you appreciate the detailed solutions?
- Were the examples useful? Did they serve as a helpful guide?
- Were you able to understand the ideas at the beginning of the chapter?
- Did this book offer good practice for you?
- Would you recommend this book to others? If so, why?

Do you believe that you found a mistake? Please email the author, Chris McMullen, at greekphysics@yahoo.com to ask about it. One of two things will happen:
- You might discover that it wasn't a mistake after all and learn why.
- You might be right, in which case the author will be grateful and future readers will benefit from the correction. Everyone is human.

ABOUT THE AUTHOR

Dr. Chris McMullen has over 20 years of experience teaching university physics in California, Oklahoma, Pennsylvania, and Louisiana. Dr. McMullen is also an author of math and science workbooks. Whether in the classroom or as a writer, Dr. McMullen loves sharing knowledge and the art of motivating and engaging students.

The author earned his Ph.D. in phenomenological high-energy physics (particle physics) from Oklahoma State University in 2002. Originally from California, Chris McMullen earned his Master's degree from California State University, Northridge, where his thesis was in the field of electron spin resonance.

As a physics teacher, Dr. McMullen observed that many students lack fluency in essential math skills. In an effort to help students of all ages and levels become fluent in mathematics, he published a series of math workbooks on fractions, long division, word problems, algebra, geometry, trigonometry, logarithms, calculus, probability, differential equations, complex numbers, and more. Dr. McMullen has also published a variety of science books, including astronomy, chemistry, and physics workbooks.

Author, Chris McMullen, Ph.D.

Essential
CALCULUS
Skills Practice Workbook
with Full Solutions
$\frac{d}{dx}\tan(5x)$
$\int \sqrt{1-x^2}\,dx$
Chris McMullen, Ph.D.

Sequences
and
Series
Calculus Practice Workbook
$\sum_{n=1}^{\infty}\frac{1}{n(n+2)}=\frac{3}{4}$
$\frac{1}{1\cdot3}+\frac{1}{2\cdot4}+\frac{1}{3\cdot5}+\frac{1}{4\cdot6}+\cdots=\frac{3}{4}$
Chris McMullen, Ph.D.

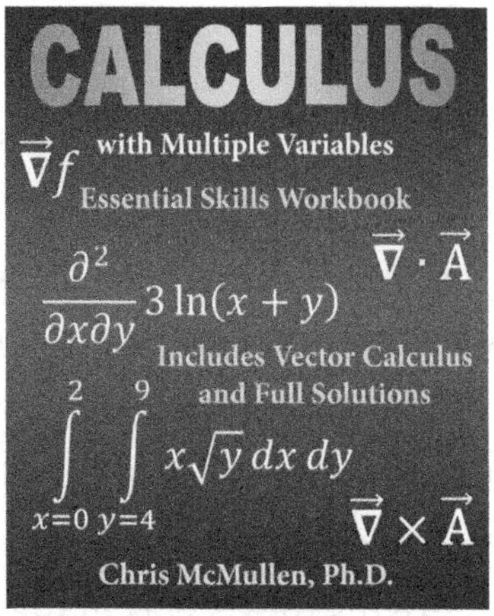
CALCULUS
$\vec{\nabla}f$
with Multiple Variables
Essential Skills Workbook
$\frac{\partial^2}{\partial x\partial y}3\ln(x+y)$
$\vec{\nabla}\cdot\vec{A}$
Includes Vector Calculus
and Full Solutions
$\int_{x=0}^{2}\int_{y=4}^{9}x\sqrt{y}\,dx\,dy$
$\vec{\nabla}\times\vec{A}$
Chris McMullen, Ph.D.

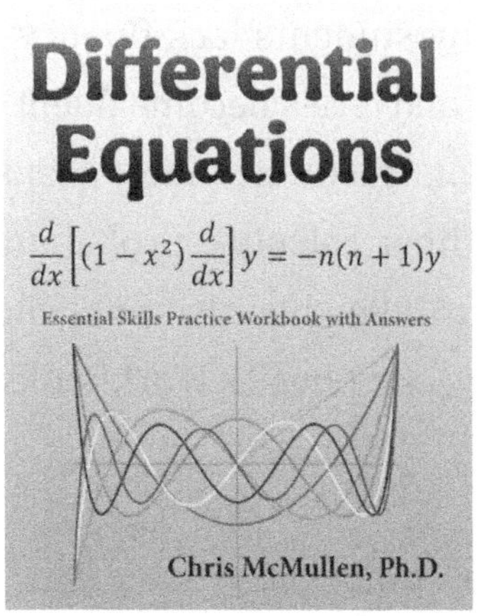
Differential
Equations
$\frac{d}{dx}\left[(1-x^2)\frac{d}{dx}\right]y=-n(n+1)y$
Essential Skills Practice Workbook with Answers
Chris McMullen, Ph.D.

ESSENTIAL
PROBABILITY
PRACTICE WORKBOOK
WITH ANSWERS
CHRIS MCMULLEN, PH.D.

50 CHALLENGING
CALCULUS
PROBLEMS
$A=\int_{x=-2}^{2}(4-x^2)\,dx$
FULLY SOLVED
Chris McMullen, Ph.D.

TRIGONOMETRY
Essentials Practice Workbook with Answers
Master Basic Trig Skills
$\cos(\alpha+\beta)=$
$\cos\alpha\cos\beta-\sin\alpha\sin\beta$
Improve Your Math Fluency Series
Chris McMullen, Ph.D.

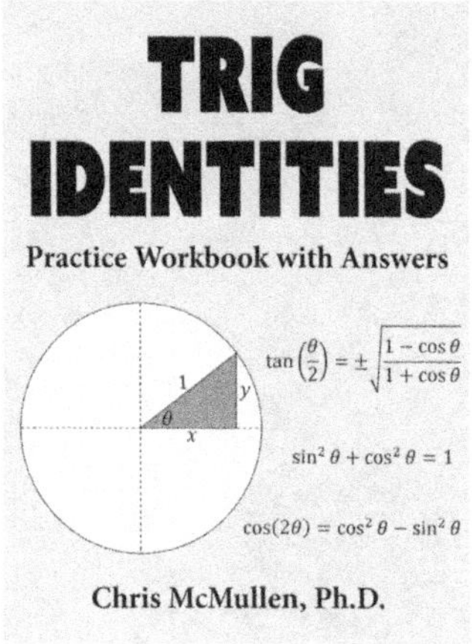
TRIG
IDENTITIES
Practice Workbook with Answers
$\tan\left(\frac{\theta}{2}\right)=\pm\sqrt{\frac{1-\cos\theta}{1+\cos\theta}}$
$\sin^2\theta+\cos^2\theta=1$
$\cos(2\theta)=\cos^2\theta-\sin^2\theta$
Chris McMullen, Ph.D.

Chris McMullen, Ph.D.
$\log_3 81 = ?$
$\log_4 32 = ?$
$\ln e = ?$
$e^0 = ?$
$\ln(e^x) = ?$
Essential Skills Practice Workbook with Answers
$2^9 = ?$
$\cosh(x-y) = ?$
LOGARITHMS
and
EXPONENTIALS

Essential Calculus-based
Vol. 1
PHYSICS
Study Guide Workbook
Volume 1: The Laws of Motion
Chris McMullen, Ph.D.
Includes Answers!

Essential Calculus-based
Vol. 2
PHYSICS
Study Guide Workbook
Volume 2: Electricity and Magnetism
Chris McMullen, Ph.D.
Includes Answers!

Essential Calculus-based
Vol. 3
PHYSICS
Study Guide Workbook
Volume 3: Waves, Fluids, Sound, Heat, and Light
Chris McMullen, Ph.D.
Includes Answers!

100 Instructive Calculus-based
PHYSICS
Examples
Volume 1: The Laws of Motion
Chris McMullen, Ph.D.
Vol. 1

100 Instructive Calculus-based
PHYSICS
Examples
Volume 2: Electricity and Magnetism
Chris McMullen, Ph.D.
Vol. 2

100 Instructive Calculus-based
PHYSICS
Examples
Volume 3: Waves, Fluids, Sound, Heat, and Light
Chris McMullen, Ph.D.
Vol. 3

Essential
MODERN
PHYSICS
Study Guide Workbook
with Full Solutions!
Chris McMullen, Ph.D.

A Guide to THERMAL PHYSICS
from the Fundamentals thru Callen-Level Equilibrium Thermodynamics
Chris McMullen, Ph.D.
Problems Have Hints & Answers

An Advanced Introduction to
CALCULUS-BASED PHYSICS
(Mechanics)
Chris McMullen, Ph.D.
Problems Have Hints & Answers

GEOMETRY
PROOFS
ESSENTIAL PRACTICE PROBLEMS
WORKBOOK WITH FULL SOLUTIONS
CHRIS MCMULLEN, PH.D.

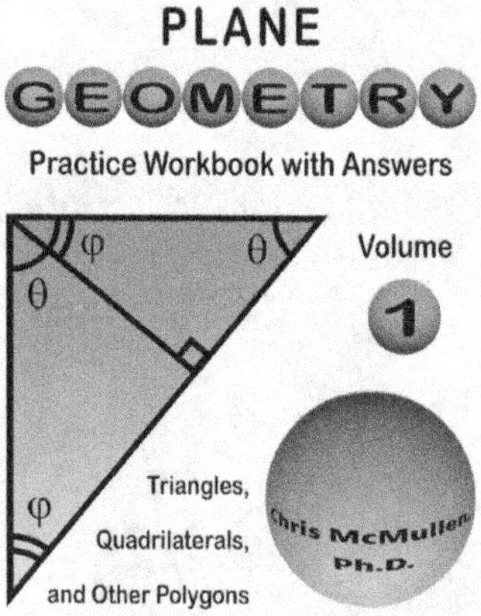
PLANE
GEOMETRY
Practice Workbook with Answers
Volume
1
Triangles,
Quadrilaterals,
and Other Polygons
Chris McMullen
Ph.D.

PLANE
GEOMETRY
Practice Workbook with Answers
Volume
2
Circles, Chords,
Secants, and Tangents
Chris McMullen
Ph.D.

COMPLEX
NUMBERS
Chris McMullen, Ph.D.

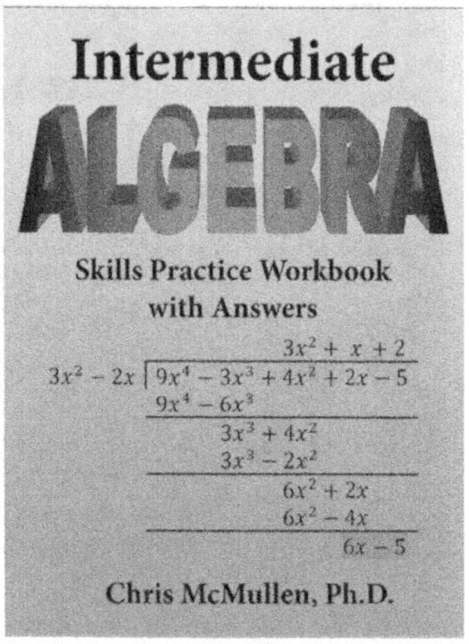
Intermediate
ALGEBRA
Skills Practice Workbook
with Answers
Chris McMullen, Ph.D.

101 Involved
ALGEBRA
Problems with Answers
Chris McMullen, Ph.D.

WHAT IS
CALCULUS?
CHRIS MCMULLEN, PH.D.

WORD
PROBLEMS
with Answers
Sam is 20 years older than Amy. 8 years ago, Sam was
twice as old as Amy. How old are Sam and Amy now?
8 years ago
now
Amy
20
28
Sam
40
48
Chris McMullen, Ph.D.

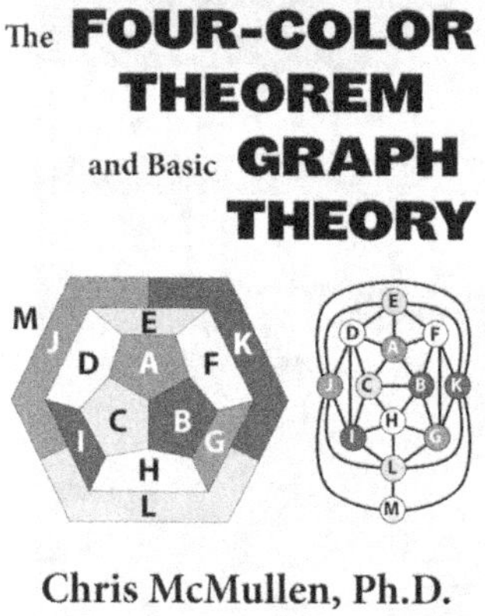
The FOUR-COLOR
THEOREM
and Basic GRAPH
THEORY
Chris McMullen, Ph.D.

An Introduction to
Basic Astronomy Concepts
(with Space Photos)
Chris McMullen, Ph.D.
Northwestern State University of Louisiana

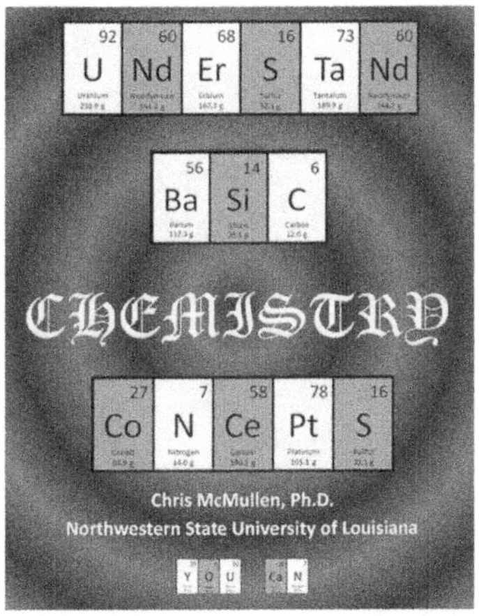
U Nd Er S Ta Nd
Ba Si C
CHEMISTRY
Co N Ce Pt S
Chris McMullen, Ph.D.
Northwestern State University of Louisiana

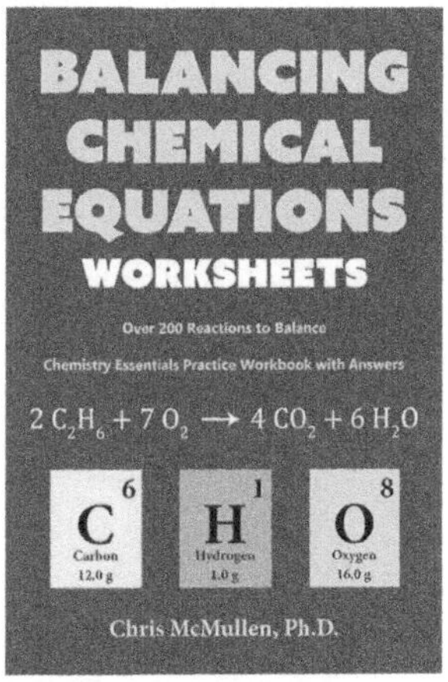
BALANCING CHEMICAL EQUATIONS
WORKSHEETS
Over 200 Reactions to Balance
Chemistry Essentials Practice Workbook with Answers
$2\,C_2H_6 + 7\,O_2 \rightarrow 4\,CO_2 + 6\,H_2O$
C Carbon 12.0 g
H Hydrogen 1.0 g
O Oxygen 16.0 g
Chris McMullen, Ph.D.

The Visual Guide to
EXTRA DIMENSIONS
Volume 1
Visualizing the Fourth Dimension, Higher-Dimensional Polytopes, and Curved Hypersurfaces
Chris McMullen, Ph.D.

V Er B Al
Re Ac Ti O N S
Word Scrambles with a Chemical Flavor
MEDIUM
Rearrange Symbols from Chemistry's Periodic Table to Unscramble the Words
S + Ni + Ge + U → GeNiUS
2 C + N + 2 I + P → PICNIC
Ti + C + Cr + P + Y → CrYPTiC
2 C + U + 2 S + Es → SUCCEsS
Chris McMullen and Carolyn Kivett
FUN FOR WORD PUZZLE FANS! NO SCIENCE NEEDED!

The FOURTH
DIMENSION

300+
MATHEMATICAL PATTERN PUZZLES
NUMBER PATTERN RECOGNITION AND REASONING
CHRIS MCMULLEN, PH.D.

PYRAMID MATH
PUZZLE CHALLENGE
CAN YOU FILL IN THE MISSING NUMBERS?
CHRIS MCMULLEN, PH.D.

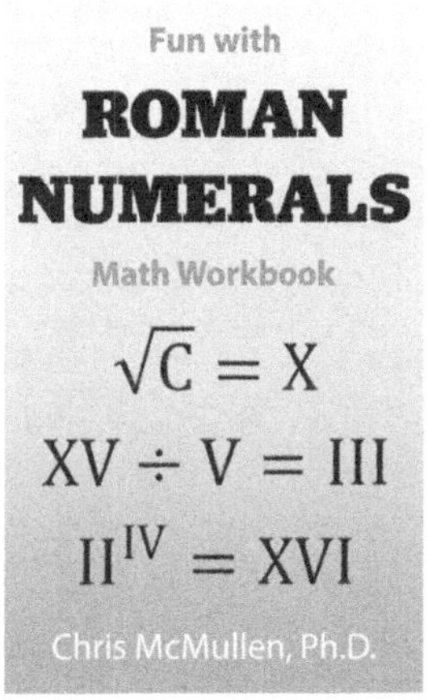
Fun with
ROMAN NUMERALS
Math Workbook
$\sqrt{C} = X$
$XV \div V = III$
$II^{IV} = XVI$
Chris McMullen, Ph.D.

Made in the USA
Coppell, TX
16 October 2025